Man and Woman

Man and Woman

An Inside Story

DONALD W. PFAFF, PhD

Professor and Head of Laboratory
Neurobiology and Behavior
The Rockefeller University
New York, NY

OXFORD

UNIVERSITY PRESS

2011

OXFORD
UNIVERSITY PRESS

Oxford University Press, Inc., publishes works that further
Oxford University's objective of excellence
in research, scholarship, and education.

Oxford   New York
Auckland   Cape Town   Dar es Salaam   Hong Kong   Karachi
Kuala Lumpur   Madrid   Melbourne   Mexico City   Nairobi
New Delhi   Shanghai   Taipei   Toronto

With offices in
Argentina   Austria   Brazil   Chile   Czech Republic   France   Greece
Guatemala   Hungary   Italy   Japan   Poland   Portugal   Singapore
South Korea   Switzerland   Thailand   Turkey   Ukraine   Vietnam

Published by Oxford University Press, Inc.
198 Madison Avenue, New York, New York 10016
www.oup.com

Oxford is a registered trademark of Oxford University Press

Library of Congress Cataloging-in-Publication Data

Pfaff, Donald W., 1939–
    Man and woman : an inside story / Donald W. Pfaff.
    p. ; cm.
    Includes bibliographical references and index.
    ISBN 978-0-19-538884-8
    1. Sex differences (Psychology) 2. Sex differences. I. Title.
    [DNLM: 1. Sexual Behavior—psychology. 2. Brain Chemistry.
3. Gender Identity. 4. Hormones. 5. Sex Characteristics. BF 692 P523m 2011]
    BF692.2.P43 2011
    155.3'3—dc22   2010014550

9 8 7 6 5 4 3 2
Printed in the United States of America
on acid-free paper

Contents

Man and Woman

ONE

What Scientists Fight About When They Fight About Sex

So many theories, so little time. Ask a lot of different scientists questions about sex differences in our brains and our behavior, and you'll get a lot of different opinions. For some, it will simply be in the X and Y chromosomes. Others will focus on hormones. And who could rule out environmental causes? Importantly, some scholars will emphasize that, to some extent, we have been required to occupy cultural "slots." Sex is a cultural construct. The linguistic habits of calling people "men" and "women" have enforced sex differences even when the biology does not dictate those differences. And such scholars might note that, among languages, the distinction between man and woman may be universal.

I'll walk on both sides of the street. On the one hand, I'll argue that sex differences have been exaggerated. But, on the other hand, where sex differences really exist, I'll provide a story of biological mechanisms for their development. For sex differences in our brains that control behaviors related to reproduction, I can tell you quite a complete story, from genes and proteins through nerve cells and neural circuits. For these mechanisms to work properly, hormones and experience—particularly experience during certain critical periods of development—interact, with

MAN

पुरुष

MANN

男

HOMBRE

ЧЕЛОВЕК

男

مرد

HOMME

איש

남자

HOMEM

άνδρας

WOMAN

स्त्री

FRAU

NŐ

女

MUJER

ЖЕНЩИНА

女

عورت

FEMME

אשה

여자

MULHER

γυναίκα

Symbols for Man and Woman in several languages. Do they signal deep biological truths about brain and behavior, mere cultural habits, or both?

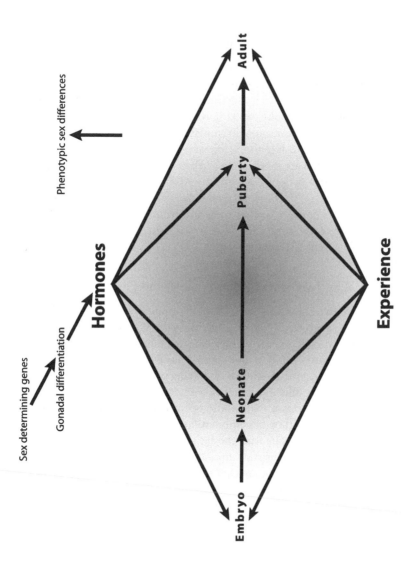

Sex determining genes

Gonadal differentiation

Hormones

Phenotypic sex differences

Embryo — **Neonate** — **Puberty** — **Adult**

Experience

My colleague at the Rockefeller University, neuroendocrinologist Bruce McEwen, has pictured a "cascade of effects leading to phenotypic sex differences." Note that experiences deriving from environmental influences can interact with the biological effects of hormones at every developmental stage. I think that newborns (neonates) and adolescents (pubertal stages) are times of particularly high sensitivity to environmental influences.

5

the result of determining sex-related behaviors in boys and girls, men and women.

The Long View

When Tu-chai-pai, god of a Native American tribe in southern California, was making the world, he first created hills, valleys, oceans and forests. He then dug into the ground to make the first people. He made the men easily, but he had much trouble making the women; he succeeded only after a number of time-consuming attempts.

Many people seem to feel the same way Tu-chai-pai must have felt at the time: that women are more complex than men—too complex, some men would say. And scores of tales and treatises, from the Biblical creation story to the bestselling book assigning the sexes to different planets, reflect the common perception that men and women differ to the point of not being made of the same stuff. Was it worth Tu-chai-pai's trouble to make men and women differently?

Why Two Sexes? What Is Sex Good For?

Why do we have two sexes, anyway? There are many theories, and scientists disagree with each other about which factors are the most important. Several answers work together to dominate a neuroscientist's thought. The first has to do with evolution; our species has evolved through natural selection of the fittest, a process in which mixing two strings of DNA from the male and female results in a much greater set of possible bodies and minds for the next generation than would occur if we had only one string of DNA. That is, the shuffling of combinations of gene copies that results from mixing the mother's copy with the father's copy of each gene causes a greater diversity of characteristics among their offspring. In turn, if the environment changes, and the species must change correspondingly to survive, some of the extreme individuals in the more variable population have a much better chance of bringing their species through the crisis. Those individuals would not even be present in a population that lacked such variability, and so the absence of sexual reproduction would lead to extinction. Putting the same point another way, the recombining of combinations of genes in sexual reproduction permits the species to react faster to changes in the environment.

The second answer has to do with genetic mutations. Most mutations are deleterious. If you have a mutation of a gene and you have only one copy of that gene, you're more likely to contract some illness, such as cancer. But if you have two—one from your mother and one from your father—and the healthy gene dominates, you're better protected from disease. So it's reasonable, in biological terms, to have more than one sex—that is, in this argument we consider the healthiness of the DNA itself. As I mentioned, most mutations are bad. But a recessive mutation will not be expressed in the case of sexual reproduction in which the other (healthy) gene copy from the sexual mate is dominant.

Finally, for the neuroscientist, the most fascinating argument for the evolution of sexual reproduction centers on the notion that sexual reproduction allows mate choice. No sex, no mate. Mate choice can happen by aggressive males competing with other males for a passive female, or through the active choice by the female of the most attractive male. In the first case, for example, when many males are pursuing one female, the scientist (or, indeed, the fans of a young female movie star!) can hardly believe that sexual pairing is left to chance. Through either male–male competition, or active female's choice, animals (or humans) benefit when males bearing characteristics that are fortunate for their immediate environment pair off with females who are just as good in that respect, thus producing babies who will compete well in the next generation. Some theorists would emphasize males who can produce resources or defend territory; others would emphasize the female's fertility and her ability to provide high quality maternal care.

For all of these reasons, sexual selection is a complex and powerful force in nature. But how are men and women really made? Why are their behaviors and feelings different from each other, when, indeed, they are really different? What is the biological truth behind widespread beliefs about differences between the male and female mind?

Books written for the general public have not handled this subject well. For example, Anne Moir and David Jessel, in *Brain Sex*, state simplistically, "Men are different from women. They are equal only in their common membership of the same species, humankind" (p. 5). In an era when women lead major corporations, fly planes, and win Nobel prizes, this statement seems absurd. They then claim, on page 88, that "Just as puberty dramatically sorts out the girls from the boys in their behavior and their social attitudes, the hormones play their part in accentuating differences in mental abilities and aptitudes." Try to tell that to the president of the Massachusetts Institute of Technology, neuroscientist Susan Hockfield, or to the president of Princeton, molecular biologist Shirley Tilghman. Further, on page 88: "We know that chemistry largely dictates the structure of our brains...." In fact, the strong effects of sex hormones on brain development are in those primitive parts of the brain that command the pituitary gland and regulate primitive sex behaviors. But those parts of the brain, present in fish and frogs and snakes, have little to do with the structures that produce higher-order brain functions, such as logic, rhetoric, and grammar.

Sometimes the problems in discussions about sexual roles and the brain have emanated from exaggeration motivated by political attitudes. In his book, *The Sexual Brain,* Simon LeVay, who states that he is a gay man, attempts to show the dominating role of biologically deterministic factors in the assumption of gender roles by men. Genetic determination of homosexuality, for example, would take from the shoulders of gay men the weight of implied responsibility for an emotional choice of sex role. Instead, genetic determination of this role would have been beyond these men's control. LeVay is all too concerned (pp. 111–120) to assign a dominating causal role for genetic factors in the determination of homosexuality.

In this book, I'll lay out a clear understanding of how genes and environment cooperate with each other in order to make a full, sexual human being. This is not the first book in recorded history to deal with this subject. Lesley Rogers, in *Sexing the Brain* (2001), has it right. She observes that, "It is far too simplistic to think that the [gene] mapping process can be extended to all types of human behavior because it is most unlikely that any particular behavior pattern depends largely on the action of a single gene or even a string of genes" (p. 48). That is, human behavior is simply not reducible to any single gene, such that behavior can be explained in isolation as a one-to-one correlation. Consider that in 1958, molecular biologists George Beadle and Edward Tatum won the Nobel prize for studying a slime mold and enunciating the "one gene, one enzyme" principle. That discovery was great, in that Beadle and Tatum showed, for the first time, that expression of a specific gene always led to a correspondingly specific implication for the biochemistry of a cell; namely, the production of a certain protein. However, we neuroscientists now see the shortcomings of their statement. From my own lab's results, I know that, even when studying very simple behaviors in very simple animals such as mice, we never, ever have a single gene devoted to one particular behavior. Now, at best, we can say that *patterns* of genes working in certain environmental contexts regulate *patterns* of behavior.

Even a highly educated medical doctor like Louann Brizendine falls short of the mark, because of oversimplification, when she tries to talk about sex differences in her 2006 book, *The Female Brain*. Unfortunately, she puts her faith in scraps of "science" with little or no validation in the literature of hard science, such as "female-specific brain changes" culminating in "earlier maturation of decision making and emotional control circuits" (p. xix). And when she talks about the difficult task a "girl's brain" faces in reacting to hormonal and social changes during teenage years, does she really think that boys pass through adolescence unscathed? The truth is that the farther away from nitty-gritty reproductive biology we get, the less we can say unequivocally about female/male differences in the brain.

Besides their tendency to oversimplify sex differences in our brains, our feelings and our behavior, some books ignore fundamental neurological and hormonal processes that produce courtship and mating behaviors, and that are exactly the same in men and women. Consider the most important cause of our excitement as we become attracted to a potential mate: arousal of our entire brain, sexual arousal. Later, I'll emphasize the process of sexual arousal, because the neuroscientific findings I explored in my book *Brain Arousal* (2006) indicated that the brain's underlying capacity to become aroused, in general, does not distinguish males from females. Sure, men have penises and women have clitori, but the mental excitement is the same. Differences from one male to another, or differences from one woman to another, are larger than the overall differences between men and women. Women simply have to be much more careful about what they do after becoming aroused, because the consequences of pregnancy, for their bodies and their lives, are so much greater than the consequences men face.

Boy or Girl?

Since antiquity, some expectant parents have gone to great lengths to try and influence the sex of their baby. Aristotle, in the fourth century BC, claimed that the chance of having a son increased with the heated passion of the man during intercourse; he advised elderly men, presumably because he thought they lacked passion, to impregnate their wives in the summer if they wanted male heirs. But probably no one went as far as certain French aristocrats went: desperate for a male heir, some of them reportedly had their left testicles surgically removed to improve the probability of having a son. This bizarre practice stemmed from the belief, which had apparently originated in the teachings of fifth-century-BC Greek philosopher Anaxagoras, that males were formed by sperm derived from the right testicle.

Of course, today's couples in the United States, and many other countries, can turn to fertility clinics that offer sperm-sorting services, which separate X from Y sperm by relying on the differences between the X and Y chromosomes, if the couple want to predetermine the sex of their child. Parents looking for more natural methods opt for folklore strategies, such as adhering to a special diet or choosing particular times for intercourse, none of which are known to consistently result in producing babies of a desired sex.

The X and Y chromosomes stand out as unique. All the other chromosomes, the nonsex chromosomes, are used for all the nonsex developmental processes. All human beings are created according to the same, general, genetic blueprint—but when it comes to determining a person's sex, there is no symmetry, no matching sets of "his" and "hers" instructions for making a male or a female. The opposite sexes do not come into being by means of symmetrically opposed biological commands, nor do these commands complement each another like two harmonious halves of a yin-yang circle. Rather, humans become male or female by following two strikingly distinct developmental paths, which overlap only for brief spells, and at certain junctions even engage in an embryonic version of a battle of the sexes, fought out in the womb. Later, I will examine how embryonic organs start off as female and then, if there are rising testosterone levels, take on male characteristics.

Brain and Body, Orchestrated

This book tells the story of how, when, and why males and females come to behave differently under circumstances where, in fact, they really do. Based on the work of scientists around the globe, it is the story of a dramatic set of events that range from tiny biochemical changes in DNA, to human responses to stress during adolescence.

This story begins with a tiny sperm swimming toward a large egg. It will penetrate the egg's membrane, and chromosomes will mix. As a result, genes will be expressed differently in fetuses that will become males, compared to fetuses that will become females. Their sex organs start out "neutral," the same in all fetuses regardless of whether they carry XX or XY sex chromosomes. But different parts of sex organs develop through time, producing ovaries and testes. At this point, you can literally *see* the female and the male developing separately.

From the testes, testosterone will flow. It will reach the brain and the brain will never again be the same. Genes will be expressed differently, between males and females, in some neurons. Some nerve cells will develop differently in males and females. Circuits will be altered. Behavior will change.

Among laboratory animals, behavioral change will manifest itself sexually. Females will display sex behaviors that control the entire process of reproduction. Males will be attracted to females by virtue of the females' pheromones and the females' courtship behaviors. Males will be aggressive toward other males, fighting over every commodity in sight. Females will be amazing as they take care of their babies, sometimes in miserable circumstances like subways and ditches.

What about men and women? So many things about us are equal, but in some extraordinary individuals we can find extraordinary abilities and disabilities. Yes, a large number of the math geniuses have been men; and verbal fluency tends to be greater for women. Disabilities, too. For example, autoimmune diseases are suffered predominately by women, while the diagnosis of autism is about four times greater in boys than in girls. Are genes, or the environment, the reason for these differences? You guessed it. Both genes and environment, and the interactions of their effects, are involved.

This book tells a story in progress. Every week the biological and psychological scientific literature enriches our knowledge of the genetic, the nervous, and the hormonal mechanisms by which males and females behave differently (when they do). I simply want to get you up to date, jargon-free. In doing so, I'll bring together genetic, hormonal, neural, and social analyses to illuminate the behaviors, the abilities and the disabilities, the desires and the diseases that differ between men and women. I'll start in the next chapter with the most concrete causes of sex differences in brain and behavior. These will have to do with DNA itself. Then, you'll find that steroid sex hormones are rather simple chemically, as well. When I reach the brain, things will get a bit more complicated. During recent years, even I have frequently been surprised by discoveries about sex differences in the brain that have shown up in the scientific literature. It will be a pleasure to read about how these genetic and hormonal actions result in normal, friendly social behaviors (Chapter 5), but another consequence will be the aggressive behaviors more typical of males (Chapter 6). It is incumbent on me, after dealing with the adolescent period, to discuss the discomforts and illnesses of brain and behavior that are markedly different between male and female, including (Chapter 10) abnormalities of sexual development. The last chapter (Chapter 11) presents a nuanced view of circumstances in which sex differences have been exaggerated, but also a clear vision of how genes and environment cooperate to cause sex differences in behavior under circumstances where they really do reliably occur.

Before wrapping up this chapter, I must offer a word about words: sex and gender. My best dictionary says that *sex* comprises "the sum of the morphological, physiological and behavioral peculiarities of living beings that subserves biparental reproduction...." and goes on to talk about genital union. In contrast, the same dictionary's entry about *gender* emphasizes linguistics—our ways of referring to people and things according to their distinguishing characteristics such as "social rank, manner of existence and sex...." A contemporary neuroscientist's use of these words resonates quite well with these dictionary definitions. As far as brain and behavior are concerned, *sex* is the simpler idea. Is the behavior or set of behaviors tightly connected to sexual reproduction? To ovulation, copulation or care for the newborn? Then they are sexual behaviors. *Gender,* much more complicated, can include sex and its sexual acts, but is by no means limited to them. People assume gender roles in society as determined not only by their hormones, but also by their overall emotional profiles and their physical and social environments. And typical gender roles can change, within a given society, in a relatively short time. When I was a little boy, *men*—and you could even say *masculine individuals*—could become airline pilots and captains of industry. Not women. At the same time, *women* would be expected to stay home and take care of the kids. Those were *feminine* gender roles. Of course, by now, things have loosened up. One thing remains the same: *gender roles* have to do with patterns of behavior that an individual chooses to assume in society, and they, in turn, influence how society treats that individual.

There is a bottom line, a clear destination where this saga will take us. As you'll see, biological influences on sex differences in brain and behavior operate at so many different levels, and they interact with environmental influences in so many different ways, that rigid, stereotyped ideas about what is and is not typical male or typical female behavior have become impossible to sustain. Gender roles are flexible, reversible and not all-or-none, especially in modern societies. At the end of this book, you will have begun to appreciate exactly where the incredible richness and variety of gender roles comes from.

An Inside Story

I can relate the saga of this book as an "inside story" because I've watched it and lived it for so long. Neuroscience is a relatively new discipline, and the extension of neuroscience to include sex hormones and behavior is even newer. Thus, I've known all the major players, and my own lab has contributed to the story for more than forty years.

Is this question of sex differences in brain and behavior really leading to an important story?

Yes, sex sells because sex matters. Sex differences enter our national and international conversations all the time, in trivial topics like sports, and in topics of the greatest importance, such as war. Consider the following headlines regarding sport:

Gender Test after a Gold-Medal Finish
Gold is Given but Dispute over Runner's Sex Intensifies
Sex Verification: More Complicated than X's and Y's

Is the champion South African athlete Caster Semenya running in the women's races legitimately? The officials seemed to be reasoning backwards, from the fact that her times were so low, to the possibility that she did not belong in a women's race. And her Italian competitor said, "These kind of people should not run with us. For me, she's not a woman. She's a man." Some of the complexities of sexual development treated below, as you will see, are entering into their arguments.

And, some headlines regarding war:

G.I. Jane Stealthily Breaks the Combat Barrier as War Evolves
Living and Fighting alongside Men, and Fitting In

"Vive la différence!"

The cartoon-like rendering by Susan Strider, graphic artist in my lab, of the ever-present question: When should we celebrate sex differences in brain and behavior, and when should we recognize how limited they are?

Active-duty servicewomen in the United States Department of Defense now comprise about 15% of total personnel, and more than 220,000 of American troops sent to Afghanistan and Iraq since 2001 have been women. Ann Dunwoody became the first woman to achieve the highest U.S. Army rank, 4-star General; and Veronica Alfaro, on her way to earning a Bronze Star, was quoted as saying "I did everything there. I gunned. I drove. I ran as a truck commander...." Debate over these issues of gender roles in the military are not likely to end soon.

Likewise in the world of business. Claire Shipman, a senior national correspondent for ABC television news, and Katty Kay, a new anchor for the BBC in Washington, in their book, *Womenomics*, celebrate an historic shift for women in the workplace. Because women are likely to have a different pattern of responsibilities in adult life than men, and because women now occupy positions of high responsibility, the natures of the work week and of the corporate ladder are likely to change.

Throughout the book, I'll try to sustain points of view that seem very different, but that are not really opposed to each other. On the one hand, I'll relate genetic and hormonal and neuronal mechanisms that underlie sex differences in brain and behavior. On the other hand, I'll urge that the applications of this work not be exaggerated; that the farther away is the topic of discussion from the nitty-gritty of reproduction, the less important the sex difference may be.

I'll start, in the next chapter, at the very simplest level, with X and Y sex chromosomes.

TWO

Chromosomes for Him and Her

Genetics isn't destiny, but a little genetic understanding will help you understand one level of explanation of how men and women come to behave differently (when, indeed, they do behave differently). And, understanding this level is so very useful because, in these primitive aspects of biology of mind, animals are very, very similar to humans. Although, in these chromosomal details you'll read about in a minute, males and females are distinctly different, when we add other levels of analysis you will be able to see where the flexibility and reversibility of gender role comes from.

A human sperm the size of the head of a pin is wending its way toward an egg 30 times its size. Among the sperm's chromosomes, the sex chromosome might be an X or it might be a Y. The egg's sex chromosome will certainly be an X.

What's the difference between an X and a Y chromosome? For one thing, the Y is much smaller. On that small Y chromosome, there are two special regions that deserve our attention. One part of the chromosome houses a gene called SRY, and this is the only place that the SRY gene lives. The geneticists Robin Lovell-Badge and Peter Goodfellow,

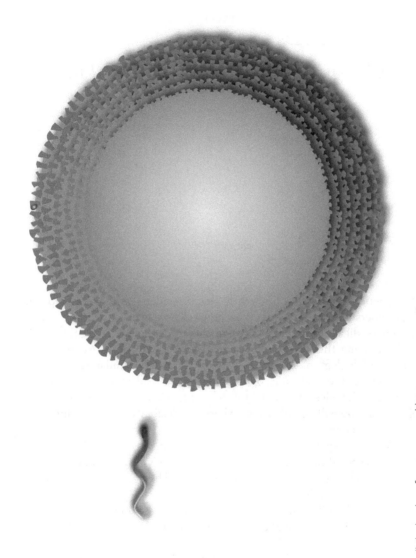

A realistic drawing of sperm approaching egg.

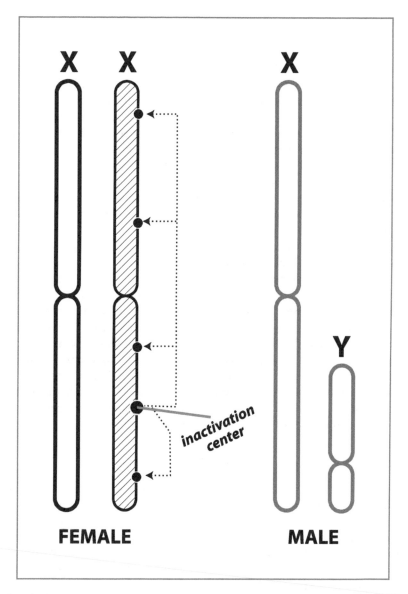

Females have two X chromosomes, each expressing many genes and duplicated on the other X. In order to control the "dose" of a given gene's product, one of the X chromosomes will have many genes inactivated. That X inactivation starts from a well defined "center" and spreads across most of that X chromosome. In contrast, males have one X and one Y chromosome. Although the Y chromosome is very small—it has shrunk progressively during four stages of evolution—parts of it are very important, because they are specific to males and indirectly affect our behavior. I'll talk more specifically about one of these, the SRY gene, later.

and their team at Mill Hill, the National Institute for Medical Research in London, discovered that this gene is the "master switch" whose expression controls the formation of a testis, and kicks off a cascade of genetic events crucial for masculine sex determination. In a human with an X and a Y chromosome, mutations in this SRY gene cause sex reversal from male to female. Another part of the Y chromosome is also specific to males. Stephen Maxson, a behavioral scientist at the University of Connecticut, collected a large amount of genetic evidence from studies with mice that showed that this second part of the Y chromosome houses a male-specific set of genes that are necessary to permit the high levels of intense aggression typical of male mice.

Talking about the importance of SRY expression for testis development also brings to mind a side point, a puzzle related to sex and the brain. Neuroscientists have been puzzled to find that there are mysterious similarities between patterns of gene expression in the testis and the brain. The most striking example of this fact comes from Ingrid Reisert and her group of neurochemists working at the University of Ulm, in Germany. She reported that the SRY gene is expressed in the brains of male mice and of men, but the behavioral consequences of this expression have not yet been proven. In the future, scientists will need to understand the biological advantage offered by certain genes being expressed in the brain, as they are in the testis. For now, however, let's switch from concentrating on the Y chromosome to concentrating on the X chromosome.

A New Trick, X Inactivation

Meanwhile, what's going on with the X chromosome? Notice first, that while the XY male has only a single X chromosome, the XX female has a double dose, two. In order to control the "dose" of products (the X genes' corresponding mRNAs and proteins) coming from expression of genes on the X chromosome, some of those genes must be *inactivated*. A striking example of the need to control the dose would be the androgen receptor, which allows testosterone to affect cells. The androgen receptor gene lives on the X chromosome, and you can probably appreciate the need not to have a double dose of androgen receptors in the female.

The British geneticist Mary Lyon, was so far ahead of her field in analyzing this X inactivation that for a while the process was called "Lyonization." A few days after conception of the fetus, one of the two

X chromosomes is rendered inactive, unable to express its genes, so that the other, active X chromosome can achieve levels of X chromosome products that are about equal to those in the male. The silencing begins at a particular point, marked by the expression of a particular "switch" gene on the X chromosome, and spreads like wildfire. Which X chromosome is inactivated in the female? This apparently is chosen randomly. A few X genes may escape inactivation. The consequences of this escape for the female's brain and behavior are not yet clear, but according to geneticists Laura Carrel and Huntington Willard, of Pennsylvania State University and Duke, more than 10% of X-linked genes show "variable patterns of inactivation" and thus suggest "a remarkable and previously unsuspected degree of expression heterogeneity among females." That is, if the X chromosome inactivation processes that Carrel and Willard analyzed were absolutely fixed and invariant, then gene expression from the X chromosomes could simply be predicted from X chromosome DNA sequences. But the real situation is more complicated. The "variable patterns of inactivation" reported by Carrel and Willard indicate that surprising combinations of expression of many genes will be seen in XX females. These surprising combinations lead to greater numbers of opportunities for individuated—not to say unique—physiological and psychological characteristics, the effects of XX gene expression in females.

Covering Up Our Genes

The DNA that comprises our genes does not sit naked in the nucleus of a cell. It is covered up by proteins that control the access to the surface of the DNA by chemicals that would turn genes on or off.

Here are two of the best established ways that access to DNA and subsequent gene expression can be regulated. First, little knots of these proteins covering the DNA can be tied to prevent access, and then *untied* so that chemicals in the cell nucleus can actually reach certain targeted genes and turn them on. The chemistry of these proteins thus offers an important means by which genes can be regulated from a wide variety of influences outside the DNA itself (see top of the figure). Second, enzymes might turn a segment of DNA *off* by adding to it very small chemical decorations called *methyl groups*—just one carbon atom joined with three hydrogen atoms (see bottom of the figure). Moreover, the DNA-covering proteins themselves can be altered chemically by adding methyl groups,

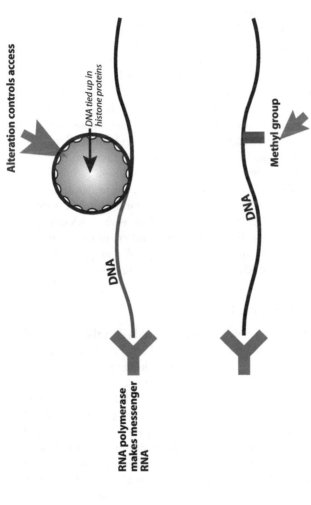

Alteration controls access

RNA polymerase
makes messenger
RNA

DNA

DNA tied up in
histone proteins

Alteration controls gene expression

DNA

Methyl group

The DNA in our cells does not sit around naked. It is covered up by proteins that control access to it, and that can impede the smooth passage of the enzyme RNA polymerase as it manufactures messenger RNA. Here are two examples of how gene expression that produces messenger RNA can be prevented, and thus regulated. TOP: The enzyme RNA polymerase, moving from left to right, will not be able to transcribe a gene which is tied up by histone proteins, and this "knot" of histone proteins also prevents access to DNA of any other proteins in the nerve cell nucleus that would facilitate gene expression. BOTTOM: RNA polymerase as it moves from left to right, making messenger RNA, will get stopped by the addition of a methyl group to the DNA. Thus, a wide variety of chemical influences on the histone proteins (TOP) or the DNA methylation (BOTTOM) can control expression of specific genes, in some cases for a very long time.

thus hindering their effectiveness. For example, the X inactivation just described proceeds via methylation of these proteins, as well as by methylation of DNA. As a result of the actions of these proteins, some genes on the X or the Y chromosome, or for that matter any other chromosome, may be permitted to be expressed. Changes of DNA-covering proteins in our nerve cells due to hormonal or environmental events can alter our behavior for the rest of our lives. For example, in laboratory animals, stress and other forms of harm imposed on the baby animal just after birth are known to influence the animal's behavior for the rest of its life, and in a small number of cases we know that methylation of DNA causally links the stress to the later behavioral changes. Such experiments add molecular detail to what we knew as the importance, for us, of the behavior exhibited by our mothers and our fathers.

Mothers and Fathers

We know that a male's Y chromosome must have come from his father. But where did his X come from? This question is important, because it turns out that some genes affect you differently depending on which parent they have come from. This phenomenon in genetics is appropriately called the "parent of origin" effect, wherein the X chromosome acts differently depending upon whether it was donated to your genetic endowment by mom or pop. Its genes have been "imprinted" by your mom's or your pop's genetic history. This is not because the sequences of DNA building blocks are different. The cause is something else.

Shirley Tilghman, now president of Princeton University, worked with molecular biological tools to help explain how imprinted genes can serve in sexual differentiation. Those same small chemical groups just mentioned, methyl groups, can be added to one of the building blocks of DNA. This chemical addition, operating in concert with the DNA sequence nearby, can turn off an entire region of DNA in such a way that the gene's expression is reduced, or even silenced. As a result, a perfectly sequenced gene coming from the father, let's say, could be silenced, while the same gene coming from the mother functions well.

Neurogeneticist Barry Keverne and his colleagues at the University of Cambridge have explored the implications of parental "imprinting" for the control of behavior. In the first place, they found that a gene may be expressed in different places in the brain according to whether it is a copy of the father's or the mother's genome. More to the point, Keverne discovered striking behavioral effects, significant alterations of behaviors of the greatest biological importance. Take the gene called *Peg3*. Keverne and his team found that Peg3 is expressed only when it comes from the father, and that mutating Peg3 causes significant impairments in maternal behaviors in mice. They were slow to build their nests, and slow to retrieve their pups to the nest. They did not assume the nursing posture typical of a normal mother. Why? Keverne feels that, among other brain effects, a reduction of oxytocin neurons in the hypothalamus—oxytocin being the neuropeptide that controls many visceral reflexes, as well as going to the mammary glands—might have accounted for the behavioral effects.

The Peg3 gene has other behavioral effects. Keverne and his team looked at the effects of experience on sex behaviors of male mice, and found that normally they become more sexually efficient with practice. The latency in time for the male to mount the female goes down, and the number of successful mounts with penile insertion goes up. This does not occur when the Peg3 gene is knocked out experimentally from a mouse's genome. Even after sexual experience, the experimental mouse's latencies to mount remain high, and the number of successful mounts stays very low. In sum, Keverne and his colleagues have proven the involvement of an imprinted gene in the regulation of adult behavior, including behavioral responses that normally are sexually differentiated.

Behold the Testis

Returning to the SRY gene story, it is timely to ask: What are the consequences of turning on a genetic XY male's SRY gene? As mentioned, its expression on the Y chromosome will turn out to be the key event in producing differences between females and males. The protein produced by SRY expression triggers a series, a "cascade" of other genes to be turned on, genes with names like SOX9 and SF1. Their composite actions will produce the development of the testes.

Early in conception, the two gonads are absolutely identical, males with females, not sexually differentiated at all. From the very beginning, each gonad has two parts—the cortex (coming from the Latin word for *rind*), and the medulla. If the baby has no Y chromosome, and no SRY to trigger that cascade of other genes, the cortex in each of the two gonads will grow and produce ovaries. If, instead, the baby has a Y chromosome, so that SRY and that cascade of genes can be expressed, the medulla in each of the two gonads will grow and produce testes. Further, in the male, a large protein called Müllerian inhibiting substance (MIS) will also be made, and will cause the cortex to degenerate, further encouraging the production not of an ovary but instead of the testes, which subsequently will secrete testosterone.

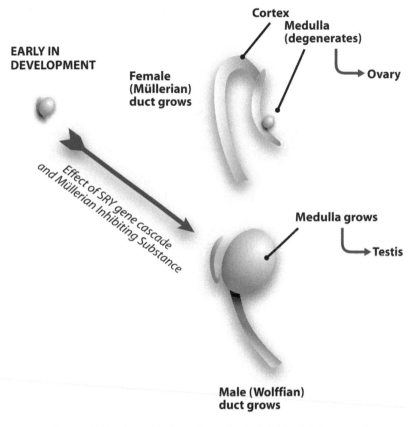

A rough sketch of how the newly formed gonad, which is identical between the sexes initially, becomes sexually differentiated. Under the influence of the SRY gene and a subsequent cascade of genetic influences, the medulla of the gonad develops into a testis. In the absence of such influences, the cortex of the gonad develops into an ovary.

A New Idea

Arthur Arnold, a professor at the University of California at Los Angeles, was facing a problem. As an expert in the sexual differentiation of brain and behavior, he was trying to account for some surprising results of experiments on the production of bird songs; specifically, testosterone-dependent songs produced by males. Here is what he was up against.

For years, the predominant theory of sexual differentiation of brain and behavior, and truckloads of solid experimental results, told us that the way in which male-specific behaviors are produced is that the testes of the baby secretes testosterone into the bloodstream, so that testosterone or its metabolites can masculinize the brain. Yet, Arthur Arnold faced a host of results, from his and other labs, that could not be explained by the predominant theory. In one experiment with male birds, testicular tissue did not cause masculine development of the brain's testosterone-dependent birdsong system. In another experiment, chemical tricks to deprive the brain of testosterone's metabolites did not much affect the masculine birdsong system, as it should have if the predominant theory was correct.

Thus, Professor Arnold had to ask the question: Could there be direct effects of sex chromosome genes on brain and behavior, independent of testosterone's actions?

Perhaps the SRY gene expression in the brain has something to do with Arthur Arnold's idea. After all, Arnold's lab subsequently discovered sex differences in the expression of other sex chromosome genes in the brain. For example, six genes known to live on the X chromosome were expressed at higher levels in the brains of females than in the brains of males. Further, Arnold's and other labs have produced several examples in which sex chromosome genes other than the obvious SRY could produce differences in brain and behavior, particularly male-typical aggressive behaviors. For example, differences in Y chromosomes explain differences in males' aggressive behaviors, compared among various strains of mice, independent of testosterone's effects on these behaviors.

Another line of evidence that plays into Arnold's idea has to do with MIS. Activated by the SRY gene, the MIS gene and its consequent protein are crucial for sex differentiation of the gonads, and might also be important in the brain. Working with Patricia Donahoe, an expert in the molecular biology of MIS and the first woman professor of surgery in the history of Harvard Medical School, the young neurologist Cristina Florea and others in my lab at Rockefeller University discovered expression of an MIS receptor in the developing mouse brain. Why would this receptor be expressed in the brain, unless this gonadally important protein was circulating to the brain and having an effect on sexual differentiation there?

Thus, Arthur Arnold's idea that genes may contribute to the sexual differentiation of brain and behavior accounts for some of the data using some behavioral measures. The full explanation of how these genes act directly on the brain, the measurement of how much behavioral power they exert, and their causal routes, all remain to be explored.

And the Nose Knows

While the testes and ovaries are developing at one end of the body, a striking development at the other end of the body is essential for all sexual function. This development has to do with Gonadotropin Releasing Hormone (GnRH), a small piece of a protein composed of only ten amino acid building blocks At one time, all nerve cells of the brain were thought to be born in the brain itself. But Marlene Schwanzel-Fukuda at Rockefeller University discovered the one, extremely important exception. While studying mouse embryos, she discovered that GnRH neurons are unique immigrants to the brain. They are actually born on the surface of the olfactory pit, and they migrate up the nose, along the bottom of the brain, and then turn into the hypothalamus. Discovered in mice, this migration is now known to occur in every vertebrate species studied, from fish through humans. GnRH is named according to its functions. In adults, long after its cells have reached the hypothalamus, GnRH is secreted onto the pituitary gland, the "master gland" hanging off the bottom of the hypothalamus, just above the roof of the mouth. There it stimulates the release of gonadotropins, large protein hormones that control all the reproductive functions of the testes in the male and the ovaries in the female. A strong and sudden dose of GnRH causes ovulation. Exciting for the neuroscientist, GnRH also promotes sex behavior. We (DP) found that administration of GnRH to female rats would significantly increase the animals' performance of lordosis behavior, the sway-back posture that comprises the primary female-typical mating behavior. That is, the same

small chemical that directs the pituitary and the gonads to secrete reproductive hormones also facilitates reproductive behavior, thus including
behavior in the classic physiologic concept, "the unity of the body."
Promulgating this concept, the great American physiologist Walter Cannon

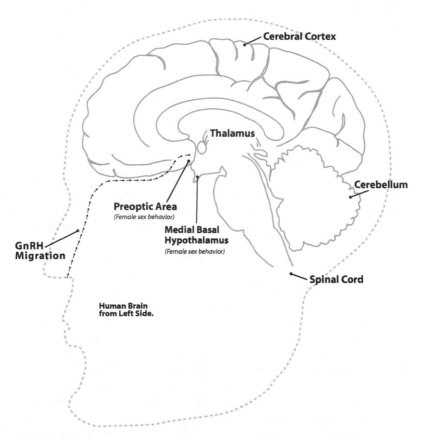

Marlene Schwanzel-Fukuda in my lab at Rockefeller University discovered that GnRH
neurons, the neurons that control all of reproduction, are not born in the brain as other
neurons are. Instead, they are born on the surface of the pheromone-sensitive portion of
the nose and, early during brain development, migrate into the preoptic area. This GnRH
neuronal migration is now understood to occur in every vertebrate species studied, from
fish to humans.

was impressed by how large numbers of organs coordinated their responses to environmental challenges. So, showing how the actions of a single molecule, GnRH, coordinates brain neurochemistry with ovulation by the ovary, and with the sexually receptive posture by the female, certainly illustrates Cannon's concept of the unity of the body.

In men, failure of GnRH neuron migration leads to an absence of reproductive functions, including the absence of sexual desire, libido. This loss of sexual maturation and libido is called Kallmann syndrome, and is due to the deletion of a gene on a specific part of the X-chromosome. Thus, it is called X-linked Kallmann's. By no means is the GnRH migration failure the only cause of Kallmann syndrome, however. The laboratory of the leading medical expert on this syndrome, William Crowley at Harvard Medical School, has just reported, based on studies of several patients, that the loss of a gene called *prokineticin2* also causes Kallmann syndrome. In the future, more such genes certainly will be discovered, but X-linked Kallmann's made an interesting point about the relations between genes and behavior in humans. These patients had no libido (1) *because* their testosterone levels were extremely low, *because* (2) gonadotropic hormones had failed to come from their pituitary glands, *because* (3) GnRH had failed to come from their hypothalamus, *because* (4) GnRH neurons had not reached the hypothalamus during development, *because* (5) of a specific neuronal migration failure, *because* (6) of gene deletion on the X-chromosome. Thus, we can prove a gene/behavior causal relationship, but the six intervening steps certainly prevent us from thinking in any simple-minded way about such a relationship.

The Story So Far

At this point we have the testes and ovaries developed, and the GnRH neurons in place in the forebrain. In males, expression of the SRY gene on the Y chromosome sets off a cascade of other genes that will help the testis to develop. In females, the very absence of that genetic cascade permits the ovaries to develop. In both sexes, neurons expressing the GnRH gene during development start their life journey in the nose, but must migrate to the forebrain. The next, very important events shaping sex differences in brain and behavior are hormonal events.

THREE

Hormones on the Brain

The flexible, multiple natures of gender roles in humans derive partly from the many levels of sexual determination—influences from the environment, and from the rest of the body, outside the brain. But in so very many cases, the crucial signal arriving at the brain, as a result of these various influences, is a hormone. In laboratory animals we can sort out cause and effect by adding or withdrawing hormones experimentally. Such experiments have led to some very clear stories.

It won't surprise you to hear that as a major consequence of the formation of testes in the male fetus, testosterone is secreted into the bloodstream. You may be surprised, however, to learn that for sexual differentiation to occur in the fetus' brain and behavior, the steroid hormone testosterone must be transformed into another steroid hormone, estradiol. This is accomplished by a special enzymatic reaction—an enzyme is a protein that catalyzes a biochemical reaction without itself being changed. In our present case, the enzyme, wrenching around a few of testosterone's atoms and bonds to form estradiol anew, provides exactly the chemical change needed. In fact, this is the only way that estradiol

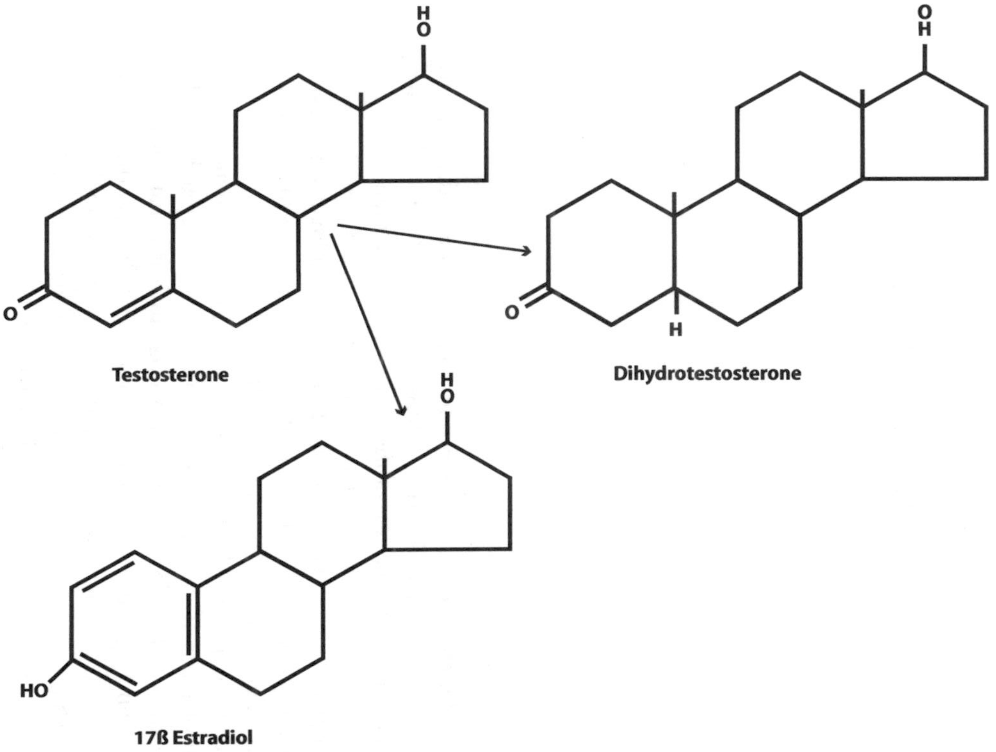

Testosterone

Dihydrotestosterone

17β Estradiol

Across a range of testosterone-sensitive brain functions and behaviors, testosterone as it affects some functions acts in its own initial chemical form, and for other functions it must have been chemically converted to the estrogen, estradiol, or the androgenic hormone, dihydrotestosterone. For example, for complete sexual differentiation of the brain, the transformation to estradiol is required. In this drawing, each intersection is a carbon atom. Carbon atoms can be joined by one bond or two bonds (the straight lines). All three steroid hormones sketched here are rigid, fat-soluble molecules, and they are quite flat (in the plane of this page of paper).

is formed. Estradiol is the most plentiful in the bloodstream of the entire class of hormones that foster feminine characters, or "estrogens."

During certain special periods in the early life of a laboratory animal or a human, testosterone secretion from the testis, and the subsequent formation of estradiol in the brain, abolish feminine hormonal rhythms and feminine behaviors. Estradiol, the estrogenic hormone, is held to be the main culprit in animals such as rats or mice. And in laboratory animals such as mice or rats, this critical period occurs during the 24 hours following birth. In a human being, a baby boy's testis will begin to secrete testosterone late in the first trimester of pregnancy, in plenty of time to affect brain development, which will continue throughout fetal life. The expert primatologist Kim Wallen, professor of psychology at Emory University in Atlanta, points out that the relative importance of the conversion of testosterone to estradiol, as compared to testosterone acting as itself, might well depend on the species studied. In the monkeys he studies, and in humans, testosterone acting in its own chemical form might be equally important for defeminizing brain and behavior as the conversion to an estrogenic hormone.

With respect to other neuroendocrine functions as well, there are differences in detail between simple lab animals and higher primates. Wallen reminded us that in higher species such as monkeys and humans, males are fully capable of showing an estradiol-induced surge of pituitary hormones of the sort that would cause ovulation. That would simply not be true in rats or mice. Whether this difference of results with primates, versus the very clear story with laboratory animals, is a slight difference of degree or a fundamental difference in principle is still not resolved.

If no sex hormones are on board during early brain development, the baby will develop with the capacity to show female sexual and maternal behaviors, and to control the pituitary gland in such a way as to regulate ovulatory cycles. In laboratory mice and rats, administration of testosterone during the critical period for brain sex differentiation (right after birth) reduces the ability of a female to show female-typical sex behaviors. A leader in these neuroendocrine studies with laboratory rats, Roger Gorski at UCLA, found that if he injected female rats with testosterone during the first five days after birth, then they could not ovulate when they grew up. They were sterile. He conversely showed that depriving XY genetic males of testosterone by removing their testes within 24 hours of

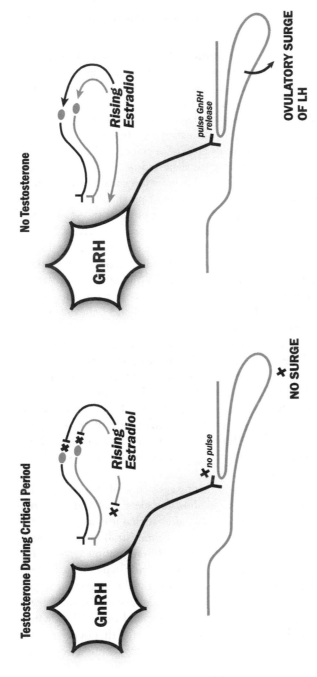

Testosterone During Critical Period

GnRH

Rising Estradiol

×i

×i ×i

×i no pulse

×

NO SURGE

No Testosterone

GnRH

Rising Estradiol

pulse GnRH release

OVULATORY SURGE OF LH

On the right, see that a normal adult female, with no testosterone having been around during the critical period for the sexual differentiation of her brain, will register the effects of rising estradiol levels directly onto her GnRH neurons, and indirectly routed through other neurons. As a result, her GnRH neurons will be able to deliver a pulse of GnRH into her pituitary gland, thus triggering the ovulatory surge of luteinizing hormone (LH). The LH circulates to the ovaries, causing ovulation. On the left, with testosterone having affected the brain during the critical period—either in the normal male, or in the female laboratory animal who was treated with testosterone at that time—estradiol does not have its normal effect on GnRH neurons. Thus, no GnRH pulse, no LH surge, no ovulation. The neuroanatomy and the dynamics of these events are shown in the next figure.

34

birth not only permitted them to take on female sex behaviors but also allowed their brains to regulate the pituitary in such a way as to produce an ovulatory pulse of pituitary hormones. They could, in fact, cause an implanted ovary to ovulate. That is, the GnRH-producing nerve cells described in the previous chapter, now located in the preoptic area just in front of the hypothalamus, send their axons to the pituitary stalk and deposit their GnRH onto the pituitary gland (the "master gland") in both males and females. But in males, including men, there is no particular rhythm—we see a pulse here, a pulse there. In females, including women, by way of contrast, the GnRH neuron has the capacity to release a whopper of a dose of GnRH onto the pituitary gland. As a result, the pituitary releases a whopper of a load of protein hormones into the blood stream. Upon receipt of these protein hormones in high quantities, and only then, the ovaries release their eggs. Of course, in women, ovulation occurs about every twenty-eight days, whereas in experimental animals like female rats and mice, it occurs about every four days.

From a behavioral point of view, I found most striking the early laboratory research by some neuroanatomists interested in hormones—Charles Phoenix, Robert Goy, and Arnold Gerall, working with the laboratory of William C.Young at the University of Kansas. They found that testosterone administered prenatally to guinea pigs not only abolished female sex behavior, but also increased male sex behavior responses to male sex hormones in adulthood. Goy quickly followed up these experiments with new studies using female rhesus monkeys. Offspring from mothers who had been administered testosterone prenatally showed more male-type sex behaviors, but also were masculinized in other aspects of behavior. The most obvious change was their increased tendency to engage in rough-and-tumble play, similar to normal males.

Thus, among the hormonal *dramatis personae*, testosterone plays a dominating role in our story. In a laboratory rat, testosterone administered even briefly between the ages of 18 to 22 days following fertilization will reduce a female's capacity to show lordosis behavior. The corresponding time in humans would be toward the end of the first trimester of pregnancy. This is defeminization of the brain, and depends, in part, on the enzymatic conversion of testosterone to estradiol in the brain itself. As compared with defeminization, masculinization of the brain is measured by the ability of the animal to produce male-typical behaviors.

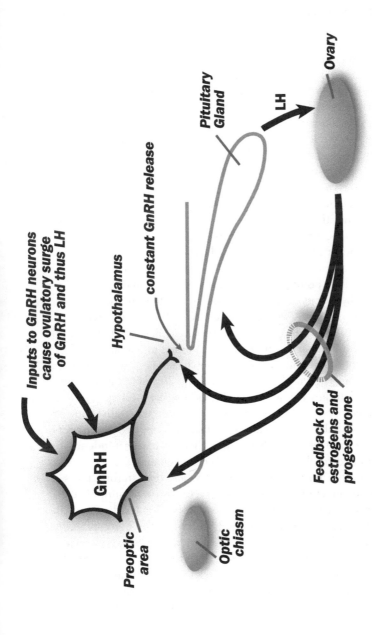

Looking at a portion of the basal forebrain from the left side, the preoptic area is just in front of the hypothalamus. The GnRH neurons in the preoptic area, just above the crossing of the optic nerves (optic chiasm), send their axons down to the stalk of the pituitary gland, which hangs off the bottom of the hypothalamus. Relatively constant release of GnRH there, thus to bathe the the pituitary, would be typical in the male. The male has no rhythm. But in the female, under the right conditions of rising estradiol levels (look back at the previous figure), a tremendous pulse of GnRH is released onto the pituitary gland, thus causing the pulse of LH that triggers ovulation by the ovaries.

Now, in addition to whatever defeminizing roles estrogens in the brain may play, this process, as well as masculinization of the brain, requires androgen receptors. Androgen receptors are large proteins that contain a special chemical site that binds "andro-genic" ("male generating") hormones such as testosterone and its metabolite, dihydrotesterone (DHT). Genetic abolition of androgen receptor function, achieved by the molecular endocrinologist Takashi Sato and his colleagues at the University of Tokyo, absolutely prevented the neonatal hormonal induction that usually would lead to the ability of an XY male, tested as an adult, to produce male sex behaviors.

Androgen receptors move back and forth between the cell's cytoplasm, just beneath the cell membrane, and the cell nucleus (the DNA-containing portion) of a cell. When they are not bound to testosterone, they reside mainly in the cytoplasm. Given testosterone, within 30-40 minutes the androgen receptors move to the cell nucleus. There, in the nucleus, when androgens such as testosterone have a cellular effect—in the brain, that might mean an effect on male sex behavior—they are depending not just on the androgen receptor itself, but also on a suite of nuclear proteins that team up with the androgen receptor as it sits on the DNA. This suite of proteins strengthen and stabilize the androgen receptor's bond to the correct DNA nucleotide sequences on androgen-responsive genes, long enough for the androgen receptor to facilitate the initiation of transcription in those androgen-dependent genes (and *only* androgen-dependent genes).

In the brain, hormones coming from the testis must regulate production of the androgen receptor, as demonstrated by the neuroanatomist Lydia DonCarlos at the Loyola College School of Medicine in Chicago, when she found that less than half the concentration of androgen receptor was found in the brains of males castrated on the day of birth, compared to males given a sham operation. Male androgen receptor levels were reduced to those of females.

This is important for brain and behavior. Kathy Olsen, now associate director of the National Science Foundation, was the first to use detailed behavioral analyses to show that male mice and rats with a genetically disabled androgen receptor, deficient by a mistake in only one nucleotide base, not only failed to ejaculate and thus inseminate females, they also suffered from a lack of sexual motivation, in that they failed even to try and mount females. They were not "feminized," in that they did not show lordosis behavior when mounted by normal males. Instead, they could not show normal levels of male-typical behavior. Later, we'll talk about these same types of androgen receptor mutations in humans.

Questions and Complexities

As is often the case in science, scholars working independently of each other in different schools happened upon the same striking finding. The endocrinologist Frederick vom Saal at the University of Missouri, and neurobiologist Robert Meisel, now at the University of Minnesota, with Lynn Clemens at Michigan State University, all found out that, for experimental lab animals, a female's behavior could be influenced by the sex of her neighboring babies when she was developing in the uterus. Focusing on the most extreme comparison, a female developing in the uterus surrounded by two males (a "2M female") will enter puberty later, will have a shorter period of fertility in life, and will be less attractive to males than "2F females." Also, a 2M female will have less female sex behavior and more masculine behavior than a 2F female—for example, she will be more aggressive, similar to genetic males. In fact, her testosterone levels as a fetus will be significantly higher if she is surrounded by two males than if she is surrounded by two females.

Thinking of the baby in the womb, we're reminded not only of hormonal influences originating with other fetuses, but also potential hormonal influences originating from the mother. If the chemical conversion of testosterone to estradiol in the brain defeminizes the female brain, how does the developing female escape the defeminizing effects of her mother's estrogens? Bruce McEwen, renowned neuroendocrinologist working at Rockefeller University solved that one. Proteins in the blood of the female *in utero* soak up maternal estrogens and prevent those hormones from reaching the brain. True, experimental, artificially large doses of estradiol could overwhelm the binding capacity of these blood proteins, but in the normal case, the developing female brain is protected from the relatively low, physiological, normal doses of maternal estrogens.

So far, we have presented a fairly simple story. In the female, in the absence of testosterone, female-typical sex behaviors like lordosis will survive, and the female's brain will remain competent to order the pituitary to deliver a surge of protein hormones into the blood to cause ovulation.

In the male, because of testosterone emanating from his newly developed testes, behaviors typical of females will not appear, and his brain cannot order the pituitary to facilitate ovulation. But two complexities emerge. First, neurochemist Charles Roselli at the Oregon Health Sciences University, and others, have proven that there are enzymes in the brain that permit local synthesis of estrogens. Not only that, but circulating testosterone affects the developing brain in such a way as to encourage estrogen synthesis there. These enzymatic actions in the brain encourage the development of male, but not female sex behavior circuits.

A second complication: we know now that during fetal development, the female's brain is protected from the defeminizing effects of her mother's estrogens because those low physiological levels of estrogens are soaked up by a blood protein. However, neurobiologists Michael Baum at Boston University, and Julie Bakker at the University of Liege in Belgium, have gotten together to produce evidence that after birth, low physiological levels of estrogens coming from the ovaries in females actually promote the development of the brain's capacity to produce female-typical behaviors. For example, if the ovaries of female rats were surgically removed on the day of birth, and then re-implanted with other ovaries until about the time of puberty, their lordosis behavior was greater than that of females who were not re-implanted with ovaries. Further, using genetic knockout mice to prevent production of estrogens in ovaries and brain led to deficiencies in female sex behavior. Bakker and Baum's evidence suggests that even though high concentrations of estrogens at certain times during development might hinder development of feminine behaviors, low concentrations of estrogens at other times during development might actually foster their development.

How?

How do sex hormones act in the brain? Some actions are remarkably fast, as when a sex hormone binds to a molecule in the membrane of the nerve cell. As a result of this binding, chemical pathways that signal from the nerve cell membrane to the rest of the cell are rapidly altered. Chains of chemical reactions, usually featuring the addition of phosphorus atoms to proteins, quickly, within seconds, cause changes in the ion channels that control nerve cell electrical excitability. Since these chemical signals, initiated in the membrane, also reach the nerve cell nucleus, another

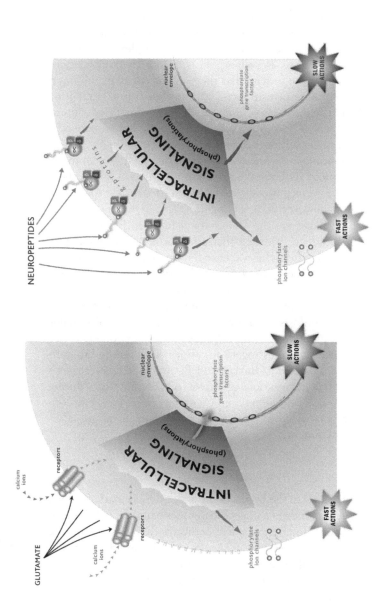

On the left, if a sex hormone binds to a molecule on the surface of a nerve cell, the "cell membrane," it can alter the rapid signaling by a neurotransmitter such as glutamate. Glutamate acts directly on ion channels (carrying small charged atoms such as calcium), with the eventual result of altering other ion channels rapidly (lower left), or slowly altering genetic transcription factors. On the right, small pieces of proteins in the nervous system called *neuropeptides* work through receptors that are much more complicated than those for simple neurotransmitters like glutamate. The neuropeptides signal through tiny complexes of sub-membrane proteins that come in triplets: alphas, betas and gammas. Again, after intracellular signaling, they can rapidly change ion channels, thus to alter nerve cell excitability or (lower right) they can causes changes in the cell nucleus that will modulate later gene expression. This is another way sex hormones can influence nerve cells.

consequence of this altered signaling is that hours later, subsequent levels of gene expression are *indirectly* changed.

Other sex hormone actions are remarkably slow. These slower actions of sex hormones in nerve cells have to do with *direct* hormone effects on gene expression. We know that during the critical period for sexual differentiation of the brain in laboratory mice, hormones coming from the male mouse pup's testis cause large numbers of genes to be expressed differently in his brain than in the female mouse pup's brain. In our lab, Jessica Mong used a technique to look at more than 11,000 genes' products during the critical period for sexual differentiation of the brain, and found significant sex differences in about five percent of the genes expressed in the hypothalamus and basal forebrain. These included genes coding for genetic transcription factors, enzymes, growth factors, and many other proteins. Among the most interesting are "connexin" proteins. Connexin proteins get together, 6 in a bunch, to form tiny tunnels or gap junctions between cells, through which electrically charged ions can move from cell to cell. Gap junctions allow electrical excitation to spread from one cell to another faster, and with greater certainty, than the regular synaptic transmission we frequently hear about. In our follow-up of Jessica Mong's experiment, neonatal male mice had greater gene expression for connexins than did neonatal females. We suspect, therefore, that among those nerve cell groups we sampled, there exists a greater capacity for the rapid spread of neuronal excitation in the male hypothalamus and basal forebrain, during the critical period for sex differentiation of the brain.

Another gene that showed up in Jessica Mong's experiments as sexually differentiated during the neonatal critical period codes for an enzyme called *prostaglandin D synthase* (PGDS). This is an important gene because the result of that enzyme's activity is prostaglandin D, a major chemical for putting an animal to sleep. In the hypothalamus and basal forebrain, neonatal males had much more messenger RNA coding for PGDS than did neonatal females. We were surprised to find that the gene actually was being expressed in the thin membrane covering these brain areas, so the gene product, prostaglandin D, must then enter the brain and act on the neurons resident in the hypothalamus and basal forebrain.

These slow, genetic actions of sex hormones require that their receptor proteins meet and bind to the hormone outside the cell nucleus, bring the hormone into the nucleus of the nerve cell, and then "find" the appropriate

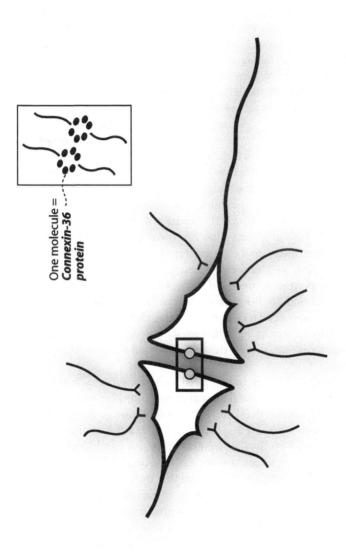

One molecule =
Connexin-36
protein

This drawing pictures a major alternative to the way we usually think about nerve cells communicating; i.e., using synapses. Drawn here are two nerve cell bodies, each receiving various synaptic inputs. They are connected by a tiny tunnel called a *gap junction*. Each of the two ends of the gap junction (pictured large, in the inset, upper right) is formed by a barrel composed of 6 molecules of the protein, connexin-36. To work right, the two cells' barrels must match up and be apposed to each other, so as to form the full tunnel between the two nerve cells. Effects of sex hormones on connexin-36 proteins forming the tunnel are important because gap junctions allow the direct passage of electrically charged molecules between neurons, thus allowing for the rapid and efficient spread of neuronal electrical excitability.

sequences of DNA that are controlling regions of hormone-sensitive genes. Our colleague at Rockefeller University, Bruce McEwen, has studied this process in the context of sex differences in brain development. He found that while androgen receptors were just beginning to be expressed in hypothalamic neurons during the critical period for brain sex differentiation in experimental animals, estrogen and progesterone receptors were already substantially expressed, though at only about half of adult levels. Thus, in the male rat for example, we expect that after testosterone from the neonatal testis has been enzymatically converted into estradiol, estrogen receptors are present in the brain in adequate numbers to serve the masculinization of the brain. Likewise, progesterone receptors are there to serve the demasculinization of the brain as we discuss below.

How exactly do the genetic actions of steroid sex hormones on nerve cells and, consequentially, on behavior come about? I can tell you from my personal experience that answering this question by doing brain research without insights that derive from research in other parts of the body would be virtually impossible. It was lucky, therefore, that our studies in nerve cells have been able to benefit mightily from the similarities between sex steroid actions in the brain, and their actions on other hormone-sensitive cells in other organs throughout the body. These slow, genomic actions require transport of the steroid sex hormone from the surface of the cell into the cell nucleus leading to events in the nucleus that will eventually lead to changes in gene expression. Here's how it works. A sex hormone gets into the cell and binds to a large protein that is specially designed to receive it. So, an estrogen would bind to an "estrogen receptor" and an androgen like testosterone would bind to the "androgen receptor." The hormone receptor escorts the sex hormone into the nerve cell's nucleus. There it diffuses through the nucleus, and if it meets a hormone-sensitive gene, it has the chance to be effective in changing gene expression. Scientists working in the laboratory of molecular biologist Jan-Ake Gustafsson, at the Karolinska Institutet in Stockholm, showed the importance of the exact structure of the steroid sex hormone, bound to its receptor, for finding its corresponding DNA sequence. Once the hormone/receptor complex finds its matching DNA sequence, a hormone-sensitive gene can be turned on. Protein-wise, this is not a solo performance. Binding of the "loaded" sex hormone receptor to the appropriate locus on DNA requires the orchestration of several

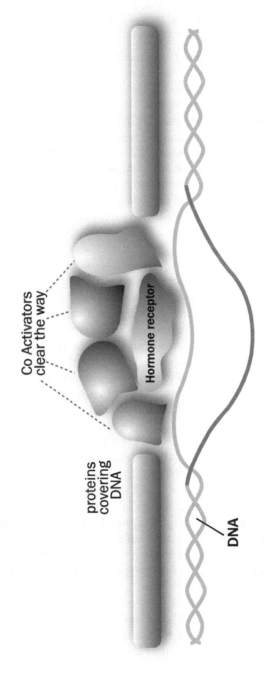

Co Activators
clear the way.

proteins
covering
DNA

Hormone receptor

DNA

Talking about hormone effects on gene expression, this drawing sketches one type of example of what can happen when sex hormones bind to their appropriate hormone receptors in nerve cells in our brains. Specialized proteins in the nerve cell nucleus, called *co-activators*, exert chemical reactions on the proteins covering the DNA. The result is to clear the way for sex hormone receptors that have bound sex hormones. As a consequence, those sex hormone receptors can grasp the DNA in the control regions of hormone-sensitive genes, and turn those genes on. Thus, sex hormones to hormone receptors, to chemical reactions on the surface of DN, to DNA access, to the turning on of hormone-sensitive genes.

proteins, called *co-activators*. Once these co-activators are themselves activated, the hormone receptor can bind to a specific sequence of DNA nucleotide bases (adenine[A]; guanine[G]; thymidine[T]; and cytosine[C]) on the portion of a gene, the "promoter," that starts transcription of the rest of the gene that codes for the messenger RNA, which, in turn, codes for a hormone-sensitive protein. For example, an estrogen receptor protein (loaded with estradiol) will bind to an "estrogen response element" that is 15 nucleotide bases long, and the androgen receptor protein (loaded with testosterone) will bind to a different sequence, the "androgen response element." As a result, these gene systems have a great deal of specificity. Out in the cytoplasm, the estrogen receptor binds estrogens and not other hormones; and, in the nucleus of the cell, the estrogen receptor binds to the estrogen response element of estrogen-sensitive genes, and not to other sequences of DNA bases. That's how estrogenic hormones turn on their own estrogen-responsive genes and not other genes.

Why do some organs respond to sex hormones and not others? Within an organ, why do some cells respond to sex hormones and not others? For both questions, the answer has to do with gene expression for the appropriate hormone receptor *and* its orchestra of co-activators. If all of those are expressed—for example, the estrogen receptor and its appropriate suite of co-activator proteins in a cell—then that cell has the capacity to respond to estrogenic hormones. If not, then that cell does not have the capacity.

In fact, using a special molecular technique to reduce expression of a particular co-activator in the neonatal hypothalamus, neuroendocrinologists Anthony Auger and Marc Tetel, working with Margaret McCarthy, an international leader in hormone/behavior research at the University of Maryland, found that this technique reduced the ability of testosterone to defeminize the hypothalamus. When males thus treated had grown up, they were able to show high levels of lordosis behavior, while the control males showed much less female-typical behavior.

The fast, membrane-initiated hormone actions do not fight with the slower, genetic actions. They *cooperate with each other*. If an estrogen is given a chance to act rapidly at a nerve cell membrane, then the subsequent genetic response to the hormone will be greater than if there was no rapid membrane-initiated action. These rapid actions also contribute to facilitating hormone-dependent behaviors. Female-typical lordosis behavior, which certainly requires genetic actions of estrogens, is brought on even better if membrane-initiated actions of estrogens occurred first. We assume that during evolution, the simpler membrane actions came first, and that the later-evolving genetic actions simply had to evolve in such a way that they would work together with the rapid actions.

Lots of Physiological Consequences

As we talk about the sexually differentiating actions of steroid hormones in the developing hypothalamus, we mainly are looking forward to the behavioral consequences described in Chapter 5. There are, however, many other consequences as well, especially for the hypothalamus' management of hormonal secretions from the pituitary gland. The most striking sex difference is that the female hypothalamus can command the pituitary to put out pulses of hormones that will cause ovulation by the ovary, whereas the male hypothalamus cannot do that. You may be surprised to hear that the liver is also sexually differentiated. Sex differences in the production of enzymes and other proteins in the liver are due to sex differences in growth hormone secretion by the pituitary. Female rats exhibit longer pulses and have altogether more growth hormone secreted. Molecular biologist Malcolm Low, at the Oregon Health Sciences University, collaborated with Marcelo Rubenstein at the University of Buenos Aires, to find that the gene coding for a hypothalamic peptide somatostatin (also known as growth hormone inhibiting hormone [GHIH] or somatotropin release inhibiting factor [SRIF]) is essential for this sex difference to appear. When that gene was knocked out, the livers of male mice were feminized. Further, the axis from the hypothalamus to the pituitary gland to the adrenal gland is not the same between females and males. Females have larger adrenal glands as a percentage of their overall body weight, exaggerated stress responses, and higher levels of circulating stress hormones. All of these examples are hormonal.

A brilliant postdoctoral scientist, Jin Zhou, working with us and with Gong Chen, a professor of biology at Pennsylvania State University, recorded the electrical activity of hypothalamic neurons, data that had been gathered during the critical period for the sexual differentiation of the rat brain. The neurons recorded are those that govern the appearance of the female sex behavior, lordosis. Remarkably, the effect of exposure to estradiol on synaptic activity recorded from these neurons was exactly the opposite between the two sexes. The estradiol exposure *increased* the frequency of activity in female neurons but *decreased* activity in male neurons. Then we looked at the flip side. What about *inhibitory* inputs, such as those from the transmitter GABA? Now the results were reversed. Estradiol exposure decreased inhibitory potentials in female neurons, but increased inhibitory potentials in male neurons. There are at least two things to say based on these results. First, the estrogen-caused changes in GABA-based inhibitory transmission could partly explain the overall increase of frequency of activity in females (and decrease in males). Second, since these neurons command estrogen-dependent female sex behavior, their increased activity following estrogen during the critical period makes a lot of sense.

Since androgenic hormones are thought to drive libido in women, in concert with ovarian estrogens (see later chapters), and these hormones are increasingly abused by adolescent girls, Leslie Henderson and her colleagues at Dartmouth Medical School looked at the electrical effects of such androgenic hormones on neurons in young female rats. Interestingly, three such androgenic hormones increased synaptic currents in hypothalamic neurons that are specialized for fostering female-typical sex behaviors, but decreased currents from neurons in the basal forebrain that are specialized for fostering male-typical sex behaviors. As with Jin Zhou's results just mentioned, the differences between brain regions might depend on the efficacy of GABA-based inhibitory transmission. The appearance of more excitatory potentials might actually follow indirectly from decreased inhibitory effects.

Progesterone

Let's talk about a steroid sex hormone we've hardly mentioned, progesterone. Because we have emphasized the role of testosterone in affecting brain development, the ability of progesterone to oppose testosterone's action comes to the fore. Behavioral neuroscientist Christine Wagner, at the State University of New York at Albany, has reported that high doses of progesterone given to neonatal male rats can reduce male sex behaviors (in males tested later as adults) and can increase female sex behaviors like lordosis. Wagner's lab has found gene expression for the progesterone receptor in neurons of the developing hypothalamus and basal forebrain, and in these same regions has detected the progesterone receptor protein. Indeed, nerve cells in the medial preoptic nucleus, a tiny cell group at the bottom of the brain just in front of the hypothalamus in the neonatal male, express the progesterone receptor much more than those of the female. Since this cell group is involved in male sex behavior, this gene expression and subsequent progesterone binding could amount to the mechanism by which high levels of neonatal progesterone could suppress male sex behavior when tested during adulthood. Likewise, males express more progesterone receptor in the ventromedial hypothalamus, a cell group responsible for facilitating female sex behavior. In turn, this could comprise the mechanism by which neonatal progesterone encourages males (tested as adults) to show lordosis. Under what circumstances might developing males be exposed to high levels of progesterone? Maternal stress. If the mother is put under considerable stress, Ingeborg Ward, endocrinologist at Villanova College, has said, then that mom's adrenal glands will pour out progesterone in amounts high enough to affect the developing male brain.

Remember Puberty?

Later we'll talk about the effects of unusual environments on boys and girls when they pass through adolescence. Here, a word about the hormonal events of puberty is in order. Although by far, the very early developmental periods just described amount to the most critically sensitive periods for sexual differentiation of the brain, sex hormones can have some influences on social behaviors all the way through puberty.

Puberty is kicked off by a massive increase in the pulsatile secretion of GnRH (aka LHRH, mentioned in Chapter 2). How does this happen? Almost always, neuroscientists look for explanations in the activities of the nerve cells we know and love. However, increasingly, another type of cell in the brain, previously thought to merely supply mechanical and metabolic support to nerve cells, has been recognized as a participant in the dynamics of signaling in the brain. This cell type is called the *glial cell*. A world-renowned neuroendocrinologist, Sergio Ojeda, and his team in Oregon, found that not only neuronal inputs to GnRH neurons in the basal forebrain, but also glial cell inputs are required for this surge of GnRH. At the very beginning of puberty, a new genetic transcriptional regulator is expressed in neurons, which enhances expression of the GnRH gene and also represses other genes that would inhibit puberty. As a result, in boys and in male experimental animals, the GnRH released from the hypothalamus will tell the pituitary to send hormones to the testes, which then will secrete dramatically higher levels of testosterone. In girls and in female experimental animals, the GnRH released from the hypothalamus will tell the pituitary to send hormones to the ovaries, which then will secrete estrogens and progesterone in such a manner as to permit ovulation. To quote Cheryl Sisk, from Michigan State University, "it is important to recognize the onset of puberty not as a gonadal event, but rather as a brain event."

The Story So Far

So, we have discussed genes, the testes and ovaries are developed, and hormones have been introduced. I have brought estrogenic, androgenic (like testosterone) and progesterone into explanations of how male and female animals behave differently, when they do. In the next chapter I'll talk about sex differences measured in structures of the brain. Knowing that in human cultures, true sex differences may account for a segment of human activity that is small but biologically significant, I am searching for the physical and chemical details of how the brain becomes sexually differentiated, in the service of producing those primitive behaviors that really are different between men and women.

FOUR

Neonatal Hormones, Brain Structure, and Brain Chemistry

It may seem a bit surprising that I can hold the position that sex differences in human cultures are limited, by and large, to patterns of behavior related to reproduction, and yet be talking about developmental hormone effects on brain structure and brain chemistry. Well, for the nervous system functions that manage the nitty-gritty of reproductive biology and associated behaviors, such as parental and aggressive behaviors, hormones and nerve cells and their interactions are absolutely required.

Once hormones have come from the testes, or have been administered experimentally, how do they affect the developing brain? First, they circulate through the bloodstream and reach the brain. There, because they are easily dissolved in fat, and the surfaces of nerve cells' membranes are covered with a fatty material, they swim into the brain freely. Then they bind to their own type of hormone receptor—testosterone to androgen receptors, estradiol to estrogen receptors, and progesterone to progesterone receptors.

According to the classical scientific literature of molecular endocrinology, the liganded receptors travel to the nerve cell nucleus, where they alter expression of specific genes. Recently, however, it has become clear that steroid hormones can also bind to receptors in the cell membrane, and affect cell activity by entirely new and different routes of action, some of them quite rapid (minutes). Within molecular endocrinology, this finding caused some difficult moments. The "nuclear guys" politically did not want to admit that there were other routes of action, that they did not have the whole story of hormone action within their own grasp. For brain research, part of the new wave of thinking derived from

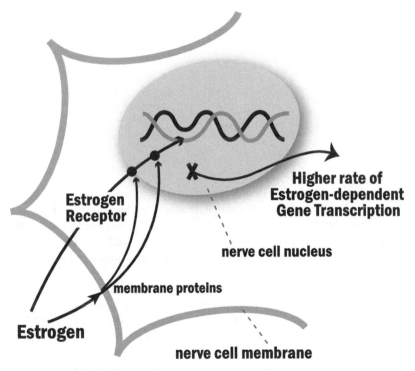

Membrane actions of estrogens and cell nuclear actions cooperate with each other. Nandini Vasudevan, working in my lab at Rockefeller University, found out that the ability of estrogens to facilitate gene expression in a neuron was greater if the estrogen administration had been preceded by an earlier effect of estrogen at the nerve cell membrane. One way of this happening, but not the only way, is if the earlier, membrane-initiated action kickstarts signals that will chemically modify (symbolized by the black dots in the drawing) the estrogen receptor on its way to the nerve cell nucleus.

the work of Nandini Vasudevan, a brilliant molecular endocrinologist, who finished her doctorate degree at the University of Bangalore, India, and entered my lab as a postdoctoral research fellow. Even though, politically, the membrane and nuclear guys were at odds, Nandini's results showed that the membrane effects of estrogens actually enhance the later, nuclear, actions. The ability of estrogens to facilitate gene transcription was greater if the cells had been exposed to an earlier, membrane action of estrogens. We theorize that in early times, steroid hormones acted in simple ways at the cell membrane, and that the more complex nuclear, transcriptional mechanisms developed later, to work alongside the more primitive mechanisms. Thus, the two modes of hormone action cooperate and synergize, even when their scientific devotees do not.

We have needed this depth of understanding of hormone actions in the brain in order to see how the hormones discussed in Chapter 3 can have such profound effects during development. That is, exposure to sex hormones, during critical periods for the sexual differentiation of the brain, makes for differences in brain structure, brain gene expression and brain chemistry.

Brain Structure

Several neuroscientists, looking for structural sex differences in the brain, concentrated on cell groups that manage sexually differentiated functions like sex behavior and ovulation. At M.I.T., I used high resolution light microscopy and found some small differences. The neuroanatomist Geoffrey Raisman, in Oxford, England, used electron microscopy and found some other small differences. But, Larry Christensen, a graduate student working with Roger Gorski at UCLA, took the prize. Using low resolution light microscopy he happened onto a cell group NOT known for managing sexually differentiated functions, but obviously different between males and females. This cell group was in the far anterior hypothalamus, just at the back of the preoptic area. It was much, much bigger in the brains of male rats than in female rats. In this case, the least detailed and laborious, the easiest scientific technique made the major advance! The neurochemical and genomic analyses I'll describe in this chapter were carried out in the brains of laboratory animals, but in Chapter 11, I'll bring in observations of sex differences in the human brain that match Larry's and Roger's discovery quite well.

Preoptic nerve cell group	♂	♀	When does difference arise in lab rat?	What function?
SDN	(large circle)	(small circle)	neonatal	♂ sex behavior
AVPV	(small circle)	(large circle)	puberty	♀ ovulation

Cheryl Sisk and Marc Breedlove at Michigan State University have pointed out striking differences between two sexually important nerve cell groups in the preoptic area. The sexually dimorphic nucleus (SDN) becomes larger in the male during the neonatal period. But the anteroventral periventricular (AVPV) nucleus becomes larger in the female during puberty.

How, during the neonatal period, does this cell group—now called the "sexually dimorphic nucleus"—get so big in males? We have four choices: more cells divide in males; more cells migrate there in males; the cells already there are larger in males; or, fewer cells die in males. Richard Simerly, an expert neuroanatomist at the University of Southern California School of Medicine, says that the first three are not the reasons. To quote Rich: "The major way sex steroid hormones alter neuronal numbers of the central nervous system is by influencing cell death." Testosterone decreases cell death in the preoptic area. Also, transformation of testosterone to estradiol causes the same sex difference in this cell group, according to the results of neuroanatomist Akira Matsumoto in Tokyo. These cells are extremely sensitive to testosterone. Mitsuhiro ("Mike") Kawata, professor of neuroanatomy in Kyoto, found out that even if a male is *simply next to* other males in the laboratory mother rat's uterus, his sexually dimorphic nucleus will be larger. This extraordinary sensitivity has an important consequence. Neurons in this region of the preoptic area have been implicated in the control of male sexual behavior.

Not all the neurons in the sexually dimorphic nucleus are chemically identical. One subdivision produces one protein, and another subgroup of neurons produces another protein, and so forth. Likewise, certain chemical details of the nerve cell group that are important for ovulation—a cell group that is farther forward in the preoptic area—has chemical peculiarities. Simerly is convinced that neurons using the neurotransmitter dopamine are especially important and, in fact, has shown an impressive gain of females over males in both cell number and the genetic capacity for dopamine synthesis in these neurons in the rat brain as it develops.

Here's a major point. When we talk about sex differences in the brain, we must, at first, talk about neurons that are present in the brain of one sex and not of the other. Once we get the neurons in place, we must ask questions about possible differences in neuronal anatomy. Are the branching inputs to the neurons, the dendrites, larger (or smaller) in the females? What about the outputs from these neurons, the axonal distributions? But then, it comes down to chemistry. Even with all the foregoing held constant between male and female brains, if one nerve cell group has chemical synthetic capacity—based on gene expression— that the other sex does not, we might get an important hormonal or behavioral sex difference.

Another interesting and important brain structural difference between males and females shows up more anterior and ventral in the preoptic area. However, in this one, the females have a larger cell group than the males. It is an important difference because of the known function of this cell group. In 1972, the young neuroendocrinologist Ei Terasawa was working at one of the world-famous centers studying hormones and the brain, in the medical school of Yokohama City University in Japan. This place burned with creative ideas. The head of the lab, Masazumi Kawakami, did not even have an office. He just stood in the middle and talked with everyone. The lab spawned international leaders in neuroendocrinology—not only Terasawa, but also the well known physiologist Yasuo Sakuma, the neuroanatomist, Yasumasa Arai, Masatoshi Yanase, and many others. Anyway, Terasawa figured out that this cell group in the anterior ventral preoptic area is necessary for ovulation. Stimulation there causes ovulation, and damage there prevents ovulation. Therefore, the female's nerve cell population and nerve cell chemistry in this small part of the preoptic area likely accounts for her ability to tell the pituitary to send hormones to the ovary and cause ovulation. Females have more cells in this cell group than males at the time of birth. But, Cheryl Sisk, neuroendocrinologist at Michigan State University, emphasizes that in laboratory animals the interesting cell group is not fully sexually differentiated during the neonatal period but instead, in her words, "arises over the course of adolescent development through a gradual increase in cell number" that is greater in females, and is associated with ability of the female hypothalamus to stimulate ovulation. That is, another huge difference between this cell group, and the sex difference discovered by Roger Gorski, is that this sex difference fully appears not during the neonatal period, but during puberty. Further, these cells are chemically different between males and females. Females have more neurons that use the neurotransmitter dopamine, which we presume to be excitatory, encouraging ovulatory mechanisms. Males have more neurons in this cell group that express opioid peptides, small molecules with opium-like effects that we'll discuss later. Since these opioid peptides tend to dampen the electrical activity of nerve cells, they may help to make sure that males cannot ovulate.

Again, the mechanisms for the sex difference in this ovulation-causing nerve cell group likely include differences in cell death. To paraphrase Simerly again: "Testosterone exposure increases cell death in this part of the preoptic area, where cell number is greater in females than in males."

These stories about sex differences in brain structure are compelling. I can mention two "twists" in the interpretations however. First, not all of the hormonal effects cease right after birth, or right after puberty. Pauline Yahr, neuroendocrinologist at the University of California at Irvine, discovered that testosterone is necessary for maintaining the volume of the sexually dimorphic nucleus in adults. This may apply more generally than just to the sexually dimorphic nucleus. Marc Breedlove, neuroanatomist at Michigan State University, found the same basic results as Yahr, but for a part of the forebrain called the amygdala. This forebrain region specializes in receiving and processing signals from biologically significant nonvolatile chemicals or pheromones.

Second, these are not the only places in the brain where we can find structural sex differences. Another cell group in the hypothalamus, the ventromedial nucleus, has different numbers and types of synapses in the male compared to the female, according to Akira Matsumoto, but we haven't figured out exactly how that difference translates into behavioral differences.

We already talked about Marc Breedlove's report about the amygdala. In fact, a lot of the other very reliable reports of sex differences have something to do with the amygdala. Catherine Wooley, neuroscientist at Northwestern University, found more excitatory synapses there in males than in females. And amygdala output targets in the forebrain, such as the septum and the prefrontal cortex, have structural differences that depend on testosterone action in the developing male brain.

All of these findings draw from research on animals. What about the human brain? First, I have to tell you that among these primitive nerve cell groups, such as those in the hypothalamus, the major features are very well conserved as we move from the lower animal brain into the human brain. Likewise, the most elementary functions of the hypothalamus, such as the female's ovulation or the male's erection and ejaculation, work quite similarly in laboratory animals and in humans. Thus, some of our own discoveries about these primitive mechanisms, such as hormone-binding neurons in the brain, and the migration of GnRH neurons from the nose into the brain, proved true from "fish to philosopher," from "mouse to Madonna."

But what about sex differences in the structure of the basal forebrain of humans? On this issue, two major figures in neuroscience faced a confrontation. Dick Swaab is a medical doctor who has been the director of the largest neuroscience research group in his country, the Netherlands, the National Institute for Brain Research. Roger Gorski received his Ph.D. at UCLA, and through his pioneering studies became the best known lecturer in the United States on subjects having to do with sex differences in brain and behavior. Swaab and his team had reported a hypothalamic cell group significantly larger in males than in females, and suggested that it is the human equivalent of the sexual dimorphic nucleus, mentioned earlier. Gorski's team, however, could not find the same thing, but did report two other hypothalamic cell groups larger in males than in females. What had happened was that the two research groups had studied brains from patients who had died at different ages, and that Roger Gorski's group had sampled brain tissue at ages when the difference was minimal. Others then followed up, and found that the reliable sex differences were due to the fact that males have a larger number of neurons, not simply larger individual cells. And Swaab later discovered that the larger number of male neurons in the cell group he had reported featured stronger expression of sex hormone receptors, both estrogen receptors and testosterone receptors. So, in the adult human brain, the magnitude, and even the detectability, of a sex difference in this nerve cell group depends on the exact age at which measurements are taken— and, from Swaab's research, we now know that the "extra" male neurons have an overload of sex hormone receptors. The exact routes by which these sex differences play into differences in human psychology still have not been determined.

Where do these intriguing nerve cell groups, that are so different between female and male, send their information? From the sexually dimorphic nucleus, fibers sweep upwards and sideways to innervate the amygdala and cell groups connected with the amygdala. They also zoom backwards to the posterior hypothalamus, and even back to the midbrain central grey. We know that some projections from the sexually dimorphic preoptic area are essential for male sex behavior, but exactly how these connections exert effects on male sex behavior we do not yet know.

The small ovulation-controlling cell group farther forward in the preoptic area is much different. Simerly tells us that many of these projections are very short, close to regions of the hypothalamus that tell the pituitary gland what to do. Most important, he has shown projections that appear directly to influence the GnRH neurons previously mentioned.

By signaling to GnRH neurons, the female-dominant preoptic nerve cell group could facilitate ovulation.

Glia

While most of us were thinking about nerve cells, neurobiologist Margaret McCarthy and her graduate student Jessica Mong rocked us back on our heels. They used a special histochemical test to look at a protein that only appears in a different kind of cell, a glial cell, in the neonatal rat hypothalamus. The etymology of "glia" comes from the Latin word for "glue," and glial cells were once thought simply to provide the mechanical support between neurons, or perhaps to shuffle nutrients and wastes back and forth. We are now beginning to understand that they participate importantly in neuron-to-neuron signaling. Thus, we all watched with interest when Mong and McCarthy found that the spread, the reach of individual glial cell's extensive processes, was much greater in neonatal male than in neonatal female hypothalamic glial cells. How do such changes in glial cells affect neuronal signals that control behavior? One possibility follows from their discovery that, coincident with this neonatal sex difference in glial cells, is a reduction of dendritic spines in these hypothalamic neurons, and so there are fewer spines for axodendritic spine synapses. That would certainly alter neuron-to-neuron signaling. As another new step in this endeavor, now McCarthy and her team had evidence of a brand new mechanism that involves more than one neuron and, likely, the glial cells that invest them both: neonatal estrogens (defeminizing agents) inducing release of the excitatory transmitter glutamate in one neuron, and thus affecting postsynaptic glutamate receptors. And, as an important result of changes in postsynaptic receptors, significant alterations of postsynaptic neuronal morphology in sexually differentiated nerve cell groups. Because glutamate transmission always involves an important "side act" featuring glial cells, these formerly neglected cells are likely in the future to be shown to play a crucial, dramatic role in the sexual differentiation of the central nervous system.

Gene Expression

One of the ways in which sex hormones in the brain work differently in females and males is in how the hormones can impact the expression of

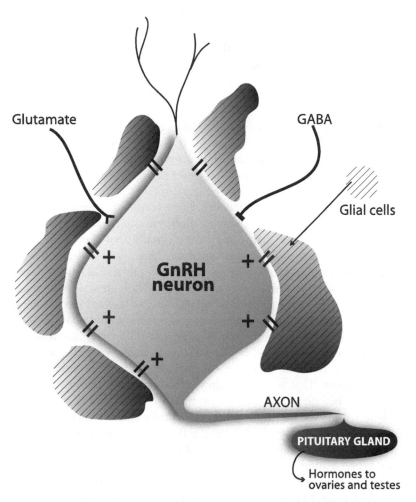

Glutamate

GABA

Glial cells

GnRH
neuron

AXON

PITUITARY GLAND

Hormones to
ovaries and testes

While almost all of our discussion so far has been about nerve cells, work in the laboratory of Margaret McCarthy and several other labs has shown the importance of another type of cell in the brain, glial cells, in neuroendocrine signaling. For example, glial cells are always implicated in signaling by the neurotransmitter, glutamate. As well, the excitability of GnRH neurons can be heightened (+) by tiny tunnels to glial cells called *gap junctions* (see the 7 = symbols sketched here). Further, Sergio Ojeda and his team at the University of Oregon have used a drawing, from which this is derived, showing the importance of molecular reactions in glial cells connected with GnRH neurons during puberty, with the effect of revving up hormone signals to the ovaries and testes. I will come back to this when I talk about puberty in Chapter 9.

genes in the nerve cells of one sex but not the other. Take the gene for the opioid peptide enkephalin, for example.

Opium yields morphine, the best painkiller we know. The brain has opioid (opium-like) chemicals, one of which is called enkephalin (from the Greek, "in the brain"). Enkephalin and other opioid peptides in the brain react to pain and stress. Gary Romano got interested in this gene system when he entered my lab. He was the type of man who disproves the bias that muscular guys are klutzes, whereas petite females are delicate and precise. Gary had the largest, strongest hands of anyone I'd seen in my lab, and yet he could carry out scientific procedures requiring care and precision as wonderfully as anyone. He is now a practicing neurologist. Gary was able to show that the female sex hormone estradiol was able to induce gene expression for enkephalin in cells of the female hypothalamus, but not the male hypothalamus. Cathy Priest, a postdoctoral researcher in my lab, further showed, in a different part of the hypothalamus, that mild stress and estrogens interact to jack up enkephalin gene transcription in females but not males. I was excited to read Gloria Hoffman's report—she is now a professor of neuroanatomy at the University of Maryland School of Medicine—that females are not simply born with more brain enkephalin. It appears in their brains during puberty, and its maintenance depends on ovarian hormones, especially estrogens. If we were to dream up ways in which adolescent girls' responses to stress differed from males', there could be worse places than the opioid gene systems to start looking.

Lots of times in biology, the important thing about a biological reaction is not just the reaction itself, *sui generis,* but what it leads to. Biochemical reactions are sometimes arranged end to end in a manner we call "cascades." As an example, we were excited when we found out that occupation of a genetic transcription factor, estrogen receptor, would cause it to go to the nerve cell nucleus and trigger the transcription of another transcription factor, the progesterone receptor (PR). What this means is that one of the hormones from the ovary, estradiol, set up the situation in the brain so that another hormone from the ovary, progesterone (P), could bind to its own receptor and change neuronal activity. Gary Romano from my lab discovered that the huge increase in PR gene expression caused by estradiol in the hypothalamus of females simply did not occur in the male hypothalamus. Roderick E.M. Scott, a scientist (and rugby player), from Glasgow—another heavily muscled but extremely precise neurobiologist—followed up Gary's work by demonstrating that

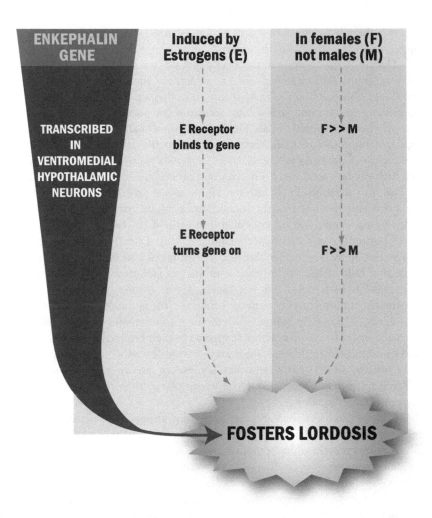

ENKEPHALIN GENE	Induced by Estrogens (E)	In females (F) not males (M)
TRANSCRIBED IN VENTROMEDIAL HYPOTHALAMIC NEURONS	E Receptor binds to gene	F >> M
	E Receptor turns gene on	F >> M

FOSTERS LORDOSIS

Sex differences in the molecular chemical steps of estrogen actions on expression of the gene for an opioid peptide, enkephalin, provide one of ways in which estrogens act on the brain to foster the female-specific sex behavior, lordosis. Lordosis is the sway-backed posture by which the female lab animal permits fertilization.

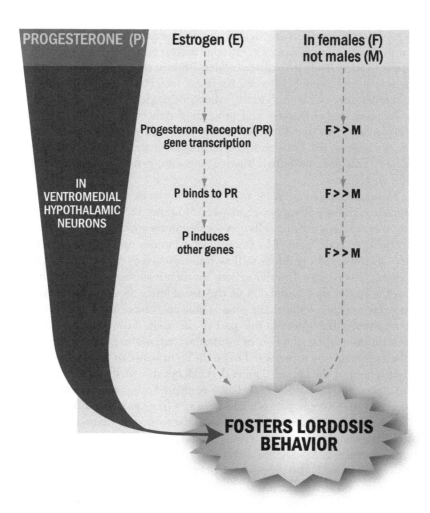

Sex differences in the molecular chemical steps of estrogen actions on expression of the gene for the progesterone receptor (PR) provide another way in which estrogens act on the brain to foster the female-specific sex behavior, lordosis. This finding excited us because even as the estrogen receptor (ER) acts as a genetic transcription factor, so does PR. So, estrogens (E) activate one transcription factor (ER) to cause synthesis of another (PR)—a multiplicative effect of the hormone E, because P often multiplies E's actions.

the most effective form of the PR is the one generated by estrogen treatment. These two hormones, estradiol and progesterone, tell the brain that the ovary is ready to ovulate. In fact, PR is itself a genetic transcription factor and, as Christopher Krebs, a molecular biologist in our Rockefeller lab showed, leads to still other gene products that are induced. We see now that estrogens' and progestins' molecular effects provide an explanation for the difference in female mating behavior; females have the genetic reactions to these hormones in the part of the hypothalamus that governs female mating behavior, genetic reactions that males do not have.

Here's another idea. We're used to the notion that nerve cells communicate with each other using synapses. However, there is also another, faster way that neighboring cells can communicate with each other. Tiny, barrel shaped gaps formed by "staves" of a protein called connexin-36 allow electrically charged ions to flow from one cell to its neighbor. Lars Westberg, a geneticist from Goteborg Sweden who was working in my lab at the Rockefeller, found that on the day of birth, the male hypothalamus expresses the connexin-36 gene significantly more than the female hypothalamus. Therefore, in this part of the brain during the period of sexual differentiation, the flow of excitation from nerve cell to nerve cell in the hypothalamus is predicted to be different between the sexes.

It is not just the hypothalamus that can boast sexually differentiated gene expression. Richard Simerly has investigated other areas of the forebrain, as well. I first met Rich when he was a free-spirited graduate student working with Roger Gorski, and with the renowned neuroanatomist Larry Swanson, from the University of Southern California. Some of Rich's experiments have focused on neural pathways related to olfaction, pathways that signal the presence of pheromones and that are different between male and female animals. He demonstrated sexually dimorphic populations of nerve cells in these pathways that express the gene for a peptide called cholecystokinin (CCK). CCK had already generated a lot of interest because of its role in conveying information about our stomach and intestines to the brain. Now, Rich has uncovered a new means of employment for this peptide. In addition, CCK expression in these pheromone-related neurons is upregulated by steroid sex hormones in both males and females, even though other genes in the same cells are not. Rich's results were specific to CCK, and we neuroscientists always think we can reason better from a result that is specific. So, chances are,

CCK promotes positive affiliative behaviors of males and females towards each other when they respond to each others' pheromones.

Likewise, neuroscientists study nerve cell groups in the preoptic area, just in front of the hypothalamus. The preoptic area contains nerve cells crucial for the female's ovulation. Take the work of Yasuo Sakuma, professor of physiology in the Nippon Medical School in Tokyo. Raised by his father, a medical doctor, in the tradition of medical research, this brilliant man made a discovery about the expression of the gene that produces a very unusual transmitter. The gene for nitric oxide synthase produces a transmitter that is actually a gas, nitric oxide. Yasuo's team found that the numbers of cells expressing the gene for nitric oxide synthase was significantly larger in females than in males. Further, estrogen treatment (followed by progesterone) reduced the number of such cells. Now, this was interesting to us because Sandra Ceccatelli, a post-doctoral researcher working in our lab at Rockefeller, showed that in the middle of the hypothalamus, where lordosis is controlled, estrogen treatment would raise the number of cells expressing this gene. In fact, nitric oxide facilitates lordosis behavior. This is the opposite result from the preoptic area—and pretty typical, given that in a number of physiological dimensions the middle hypothalamus and the preoptic area do fight with each other. Indeed, while the middle of the hypothalamus is essential for female sex behavior, the preoptic area is essential for male sex behavior.

Sometimes not just individual genes are different but *combinations* of genes get our attention. Neurobiologist Nino Devidze got her first scientific training in Tblisi, Georgia, but now has been in my laboratory at the Rockefeller University for many years. She used a technique in which she attached a tiny pipette against the nerve cell membrane, broke through that membrane, and removed the cell's contents, including the entire messenger RNA population. The most striking sex difference was not for individual messenger RNA, but for patterns of co-expression of RNAs. That is, the way estrogen receptor and oxytocin receptor gene-expressing neurons matched up with expression of signaling enzymes was significantly different between male and female hypothalamic neurons. This combination is important in behavioral terms because, when manufactured at high levels under the influence of estrogenic hormones signaling through the estrogen receptor, the oxytocin receptor allows oxytocin to foster friendly behaviors, especially in females.

We're just at the beginning of this story of genes that are expressed differently in the brains of males and females. Recently we used a technique that allows us to compare expression of more than eleven thousand genes, and found that large numbers expressed in amounts that were different between the sexes. Among them, Anthony Auger and Marc Tetel, then working with Margaret McCarthy at the University of Maryland, would emphasize sex differences in hormone receptor co-activators. Their work is of the highest importance for the regulation of behavior because these accessory proteins, co-activators, help to determine the level of sensitivity to hormones in brain tissue. Others would emphasize genes coding for peptides that act on the GnRH neurons themselves, thus influencing ovulation as it occurs in the female but not the male. Stay tuned, because this subject will develop bigtime, over the next few years.

Brain Chemistry

The most ancient neurotransmitters in the brain are glutamate and GABA. Regarding GABA, Margaret McCarthy surprised everybody. We usually think of GABA as an inhibitory neurotransmitter, because when it opens its channel in a nerve cell membrane in the adult brain, it allows a negatively charged electrical ion, chloride, to enter the cell from outside the cell. That brings the nerve cell's membrane farther away from the voltage it needs to begin firing important electrical signals that we call "action potentials." The chloride flows into the cell when GABA opens its channel in the nerve cell membrane because the concentration of chloride outside the cell is *higher* than the concentration inside. So, the chloride flows from a place of higher chloride concentration (outside the cell) to a place of lower concentration (inside).

But what Peg McCarthy pointed out was that in the neonatal hypothalamus, quite the opposite of what we expected from studies in the adult brain, the chloride concentration outside the cell is actually *lower* than the concentration inside. As a result, when GABA opens its channel, chloride flows out and the cell becomes more excitable. McCarthy has offered us the vision that neonatal sex differences in GABA signaling— excitation in males but inhibition in females—likely have to do with sex differences in transporter proteins that push chloride back and forth across the nerve cell membrane, thus to create that inside/high versus outside/low difference in chloride concentration.

Not only that—a team in McCarthy's lab, and Jin Zhou from my lab, also showed that estrogen administration enhances excitatory GABA signaling in neonatal hypothalamic neurons. We found opposite trends in our electrical recording experiments between hypothalamic neurons from females and males. In neurons from females, estrogen administration increased electrical activity, but in neurons from males the same treatment decreased electrical activity. Do we know exactly how, step by step, these biochemical and biophysical results lead inevitably to the male's inability to ovulate and to show lordosis behavior? Not yet. Instead, knowing that McCarthy's lab has been dealing in a new way with the neurotransmitter GABA, which is the dominant inhibitory transmitter in the (adult) brain, these patterns of surprising phenomena and sex differences in the neonatal brain will have to play an important part in our final understanding of how sex differences arise in the brain.

GABA is not the only transmitter system to exhibit sex differences. Robert Rubin, now chief of psychiatry at the Veteran's Administration Hospital associated with UCLA, has emphasized that functions of neural systems using acetylcholine as a transmitter appear to be more responsive to stress in females than in males. And, there are sex differences related to serotonin: Heather Patisaul, an endocrinologist at North Carolina State University, found that males had more serotonin fibers reaching the ventromedial hypothalamus, and Bruce McEwen, my colleague at Rockefeller University, found greater amounts of serotonin binding to nerve cell membranes in the preoptic area of the male compared to the female. These and other labs across the world have thus proven that in chemical terms, there is a solid, broad basis for sex differences in behavior, when, in fact, those sex differences really exist.

Sex differences are not limited to classic transmitters. Geert DeVries, born in Columbus, Ohio, educated in the Netherlands, and now a professor at the University of Massachusetts, is an expert in the neurobiology of the nine-amino acid peptide, vasopressin, a neuropeptide that preserves our fluid balance in the body. For example, if we are wounded and bleeding, vasopressin does tricks to ensure that we do not lose any more body fluids than would be unavoidable. Important for us, DeVries found that, in basal forebrain cell groups with close connections to the amygdala and preoptic area, males have at least twice as many vasopressin-expressing neurons as females. As a result, they also have more nerve cell axons signaling to other brain regions using vasopressin as the signal.

The functioning of the system is interesting, as well. Vasopressin responses to stress are different between the sexes. Stafford Lightman, a medical doctor who does high level neuroendocrine research at the University of Bristol Medical School in England, found out that in females, forced immobilization, which is quite stressful, caused a remarkable increase in the release of vasopressin into the blood. This is simply not seen in males. In terms of behavioral importance, DeVries and others think that testosterone effects on some of these vasopressin neurons are essential for the promotion of aggressive behaviors.

The Story So Far

By this point you can see that, despite the fact that only the biologically-based, more primitive behaviors related to reproduction are really convincingly sexually differentiated in humans, we nevertheless have found out a lot about sex differences in animal brains. But who cares about these effects of neonatal hormones on brain structure and chemistry? Everybody will, because in the next four chapters we'll explore their implications for animal behavior—both sexual and nonsexual behaviors—and their implications for human psychology.

FIVE

Mating and Parenting

We've come a long way in this saga, a story of sex chromosomes, developing sex organs, sex hormones early in development, and their effects on the developing brain. But for a neuroscientist, the bottom line is always the behavioral result. In the next few chapters I will explain in greater detail five major domains of behavior showing sex differences that result from the genetic, hormonal and neurochemical influences surveyed in Chapters 1 through 4. These are sex behaviors, parental behaviors, aggression, friendly prosocial behaviors, and responses to pain. Here, I'll show especially how biology explains sex differences in animal behavior, although as the discussion evolves, I'll pay increasing attention to very similar sex differences in humans.

Mating Behaviors

Let's start with sex, because many biologists consider sexual behavior the starting point, the fundamental building block for *all* social behaviors. These are the behaviors that determine what genes will be passed on to the next generation. They will be the focus of evolutionary pressures

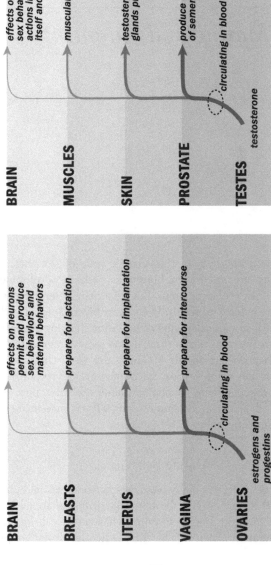

IN THE FEMALE

BRAIN — *effects on neurons permit and produce sex behaviors and maternal behaviors*

BREASTS — *prepare for lactation*

UTERUS — *prepare for implantation*

VAGINA — *prepare for intercourse*

circulating in blood

OVARIES — *estrogens and progestins*

IN THE MALE

BRAIN — *effects on neurons produce sex behaviors and muscular actions in courtship, mating itself and aggression.*

MUSCLES — *muscular growth for strength*

SKIN — *testosterone-dependent glands produce pheromones*

PROSTATE — *produce components of semen*

circulating in blood

TESTES — *testosterone*

Sex hormones are chemical signals that coordinate activities among different organs throughout the body. In the female (left page), estrogens and progestins (that is, progesterone and its metabolites) help to synchronize sex and maternal behaviors with other bodily preparations for reproduction and motherhood. In the male (right page), testosterone and its metabolites signal the brain to produce aggression (maintain a territory upon which to mate), to compete with other males and court the female, and to mate.

70

due to natural selection. Besides, given that we have enough to eat and drink, are there many things we care about more during our waking hours than sex?

At the root of these behaviors are hormones. Elizabeth Adkins-Regan, in her book, *Hormones and Animal Social Behavior*, reminds us that "hormones are coordinators" (p. 3); "they coordinate behavioral and physiological sequences over time." "They help adjust behavior to circumstances and contexts: physical, social and developmental." Males have relatively constant sex hormone levels, while females during their reproductive years almost always have fluctuating levels of sex hormones.

But in both males and females, hormone effects on behavior can be heightened either by raising hormone levels or by raising nerve cell circuitry sensitivity to a hormone. The following figure (left panel) illustrates this, and also shows a perspective that my graduate student, Sandra Cottingham, and I dreamed up several years ago (right panel). Groups of sex hormone sensitive neurons tend to project to other groups of sex hormone sensitive neurons. Therefore, as signals pass through the circuitry pathways formed by these neurons, the hormone effect may be multiplied several times over.

The man who opened up the systematic study of sex hormone effects on behavior was a twentieth-century scientist named Frank Beach. He was a hard-drinking, hard-thinking professor at the University of California at Berkeley, and some people said that if he'd lived longer he might have won a Nobel prize. The late great Professor Beach made it clear: More testosterone, more sex and faster sex.

Here's how this works: Testosterone is secreted from the testes, circulates in the blood, and enters the brain. Because it is a fat-soluble hormone and the brain is full of fat (the "lipid" membranes of nerve cells and glial cells), it goes all over the place in the brain. But, as mentioned, some cells have specialized proteins that bind testosterone and bring it into the cell nucleus, thus to alter transcription of many genes. These specialized proteins are called *androgen receptors* (from the Latin, *generating android* or masculine characteristics). The highest concentrations of cells expressing these androgen receptors are in the hypothalamus, the preoptic area, and the phylogenetically oldest parts of the forebrain. As Ruth Wood, now a professor of psychology at the University of Southern California, showed, it follows that putting testosterone directly into the preoptic area when it is absent throughout the rest of the body, following

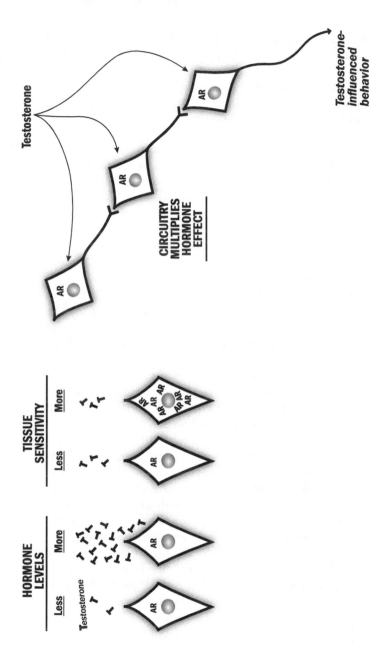

Hormone effects on behavior can be strengthened by raising hormone levels (left page, here illustrated by more testosterone, T, in the blood) and/or by raising tissue sensitivity to a constant hormone level, for example (left page) by manufacturing more androgen receptor (AR) in each nerve cell. Another way to understand how tissue sensitivity can be elevated (right page) is to notice, as Sandra Cottingham in my lab did, that hormone-sensitive nerve cell groups tend to send projections to other hormone-sensitive cell groups, offering the possibility that hormone effects on the entire circuit are multiplied. In the drawing, for example, if testosterone doubled each cell's signaling, the effect on the 3-neuron circuit sketched would be 2X2X2 = 8-fold.

castration, turns on male sex behavior. In humans, sexual desire rises as the testosterone surges of puberty arrive. For eunuchs, quite the opposite— no testosterone, no libido. That's why, in ancient Egypt, eunuchs were allowed amongst the harem.

Thus, especially important for producing male sex behavior are the androgen-binding cells in that nerve cell region just in front of the hypo-thalamus, the preoptic area. Destroying these cells destroys male sex behavior. This devastating loss of male sex behavior is not necessarily because the males lose all sexual desire for females. The loss of male mating behavior comes about because these preoptic area neurons, in particular, are responsible for controlling the intricate autonomic nervous system mechanisms that permit erection and ejaculation. Autonomic nervous systems are usually divided into two parts, sympathetic and para-sympathetic, and these are usually portrayed as opposing each other. But for producing erection and ejaculation, they are orchestrated to work together in a nuanced pattern that functions through time and depends upon tactile stimulation. The destruction of androgen-binding cells and other nerve cells in the preoptic area disrupts this orchestration.

Elaine Hull, now a professor of psychology at Florida State University in Tallahassee, has gotten some of her best results studying the neuro-chemical inputs to these neurons in the preoptic area that are so very necessary for male sex behavior. She found that nerve cell terminals releasing the neurotransmitter dopamine in the preoptic area greatly increased male copulatory behavior, even in male rats with low levels of testosterone. This makes a lot of sense, because the neurotransmitter dopamine is responsible for promoting directed motor acts toward salient stimuli. And, believe me, for the male wanting to mate, an attractive female is a salient stimulus. Not only that—testosterone acting in the preoptic area encourages dopamine release from synaptic endings there, thus facilitating male copulatory acts. According to Hull's results, the steroid sex hormone testosterone, and the neurotransmitter dopamine, work hand in hand amongst these cells in the preoptic area to permit male sex behavior. So, putting this all together, males that have normal levels of testosterone have a release of the arousing transmitter dopamine in the preoptic area as soon as they detect a receptive female. That dopamine energizes their motivation and activates their behavior. When a male has low levels of testosterone, dopamine is also low, but microin-jecting dopamine onto the neurons in the preoptic area will restore sexual arousal and sexual behavior.

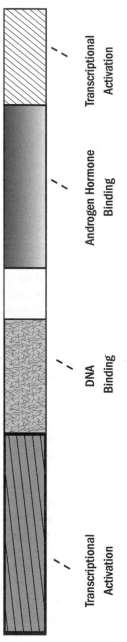

Transcriptional Activation

DNA Binding

Androgen Hormone Binding

Transcriptional Activation

Androgen receptors are large proteins, about 900 amino acids, that bind testosterone or dihydrotestosterone in a tiny enfolded pit within the androgen binding domain, and then go to the nerve cell nucleus, where the entire protein can bind to the controlling regions of androgen-sensitive genes by virtue of chemical recognition between the DNA binding domain and the appropriate DNA itself. Then, the chemical recognition of the two transcriptional activation domains at the two ends of the androgen receptor protein, with other specialized proteins in the cell nucleus, kickstart transcription, the expression of androgen-sensitive genes.

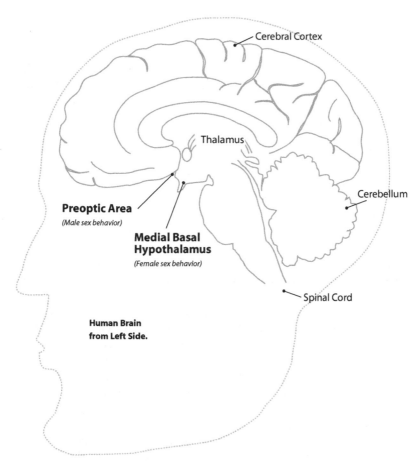

The nerve cell groups most important for producing male sex behaviors and female sex behaviors are very close to each other. They are pictured here in the human brain, but the following two statements are true for a wide variety of vertebrate species. Destroying the preoptic area greatly reduces male sex behavior, almost to zero. Destroying the medial basal hypothalamus greatly reduces female sex behavior, almost to zero.

Interestingly, Jacques Balthazart, an internationally recognized authority on neuroendocrinology at the University of Liege in Belgium, and Gregory Ball, a professor of neurobiology at Johns Hopkins University, have pointed out that these preoptic neurons are much more important for some aspects of male sexual performance than others. Destroying these neurons absolutely wipes out the male's ability to penetrate and to ejaculate, but male rats with such preoptic lesions will still pursue females and attempt to mount, and they will still learn various responses in order to gain access to females. Thus, they have sexual desire, but can not "pull the trigger."

Not all of the effects of testosterone in the brain are chemical. Some are electrical. I reported that testosterone could heighten the electrical responses of preoptic neurons to odors, a finding that might explain how male animals with testosterone show heightened interest in smells from females. Neurophysiologist Keith Kendrick, at the University of Cambridge in England, followed this up with a more discriminating study that proved the testosterone effect was directly upon the odor-signaling pathways that project into the preoptic area. Kendrick further proved that it took about the same amount of time, 5 days, for the electrical changes and the sex behavior increases to take hold, following testosterone treatment.

The genetic determinants of male sex behavior are especially interesting to explore. Obviously, the gene for the androgen receptor itself plays a part. But remember that testosterone is also converted into its daughter hormone, estradiol. And estradiol is received in the brain by not one but two receptors that are the products of a gene duplication: estrogen receptor alpha, and estrogen beta. Now things get tricky. In experimental animals like mice, either estrogen receptor alpha or estrogen receptor beta, expressed in the brain, can support the preliminary aspects of male sex behavior, like chasing the female and mounting her. But for ejaculation, the estrogen receptor alpha gene has the hammer. Knock out that gene, and the result will be no ejaculation at all. Also, damage the androgen receptor gene and you have a behaviorally demasculinized individual. The bottom line is that a complete male depends on the genes both for his androgen receptors and for his estrogen receptors. And, sometimes, those two types of receptors are expressed in the same nerve cells.

Everybody these days talks about interactions between genes and environment in the determination of behavior (I'll talk about this more in the final chapter). Clearly, our genetic inheritance influences some of our abilities and susceptibilities. But could anyone gainsay the immediate environmental effects on how "hot" a guy feels for the female who is at the center of his attention? Elaine Hull and her colleagues have documented the importance of an arousing neurochemical, dopamine, arriving at neurons in the preoptic area for stimulating male sex behavior in laboratory animals—the rapid and avid approach by the male to the female once he has received sensory signals from her. Well, stimuli from the female is a salient part of the environment for the male. And, those preoptic neurons are getting their gene expression changed by the actions of testosterone and its metabolites. So there you have it: a specific example of how a gene/environment interaction can work.

Of course, the most important aspect of the environment for a sexually motivated man is the nature of the female who is desired. In addition to looking at sex hormone effects, the famous Berkeley professor Frank Beach emphasized the role of the stimulus environments that influence sex behaviors. One of his favorite points was that many males prefer variety in the females with whom they mated. Their vigor would remain high if (and only if) they could switch from one female to another. Thus, the "Coolidge principle." An early twentieth-century American president was visiting a farm. He observed that roosters would get exhausted from mating, and the farmer would then switch them to another hen. He asked the farmer why he did that. The farmer said, "For the male, switching from female to female restores sexual vigor." President Coolidge replied, "Tell that to Mrs. Coolidge!" So, the novelty of a female potential partner is a part of the environment that makes a lot of difference.

As the Coolidge story reminds us in a semi-comic fashion, men historically were recognized more widely than women to be "looking around" for a variety of sex partners. But other men are absolutely the opposite— they have little sexual desire. Remember the nose-to-brain story about GnRH neurons I told you a few pages ago. All of male sexual feelings depend on the signals from the pituitary that tell the testes to pour out testosterone. And if the GnRH neurons failed in their journey during development from the nose into the brain, the GnRH would not be there to tell the pituitary to send out those signals. Guys in whom that GnRH nerve cell migration has failed will have no libido.

These different attitudes of males and females toward their sex life and its variety make a lot of sense because of their differential investment in the reproductive process. The male, theoretically, can inseminate and leave, while the female's biology will be tied up with the difficulties of pregnancy and baby care for a considerable period of time—in the case of laboratory rats, for a large portion of their adult lives.

Now, let's take a close look at female sexual behavior. My lab made fast progress with this subject because of an early strategic choice. That is, I was struck by the need to start with a relatively simple scientific problem, because I knew that many behaviors of more complex animals, vertebrates, would not quickly yield their scientific secrets to however detailed an analysis. So I picked the simplest mammalian behavior I could think of, and over the years made it the first whose mechanisms had been analyzed. The primary female mating behavior among four-footed mammals, lordosis, appealed to me as a strategic scientific subject for study because it is simply triggered by tactile stimuli from the male, and only requires the female to stand still and go into a swayback posture. This behavior by the female controls reproduction. It is absolutely required if fertilization by the male is to be permitted. Moreover, lordosis behavior is controlled by hormone action on the brain.

Work in my lab unraveled the simplest, most straightforward mechanisms that produce this sexual behavior. Briefly put, the cutaneous signals from the mounting by the male ascend the spinal cord of the female to reach her hindbrain and then her midbrain. There, nerve cells receive a sex-hormone-influenced signal from the ventromedial hypothalamus. If the female has received adequate doses of estrogens and progesterone, that signal from the hypothalamus says "Go, Mate, Do Lordosis Behavior." If not, the signal is "Resist, Kick, Flee the Male."

How do those ventromedial hypothalamic neurons manage to send a different signal in females than in males? The answer was discovered by Gary Romano in my lab at Rockefeller University. Romano discovered that estrogens could induce the transcription of genes in the hypothalamic neurons of females—for example, the genes coding for the progesterone receptor and for an opioid peptide (see figures above)—but the same estrogen treatments could not induce the transcription in the hypothalamic neurons of males. Romano's work showed, for the first time, how expression of a specific gene in a specific part of the brain fosters a particular behavior.

What kinds of schedules of hormone administration can trigger these changes in brain and behavior? A person might ordinarily think that for hormone administration to be effective in the brain, or in any other organ, a longer and stronger administration is always better. But that is not true.

Many years ago when I visited the campus of the estrogen expert Jack Gorski, at the University of Wisconsin, I saw that he could divide that long and strong administration into two very short hormone administrations, a "trigger phase" and a "booster phase." Evidently, that strategy of short "pulses" of hormone administration could be very important for medical treatment of women, because they could derive beneficial effects from very brief hormonal treatments. In addition, those two phases, or two pulses of hormone treatment, could be used to analyze brain mechanisms of hormone action in humans. Quickly, we applied Jack Gorski's two-pulse protocol to the rat brain and found that when estradiol was applied in these two pulses, it also promoted female sex behavior.

Lee-Ming Kow, my long-term neurophysiological colleague at Rockefeller University, then used the two-pulse estradiol treatment and added to Gary's story. He showed that in the first pulse, actions of estrogens at the hypothalamic nerve cell membrane "set up" the genetic situation so that estrogenic effects on genes, resulting from the second pulse, could later drive female sex behavior.

The sex hormone dependence of lordosis is the same as that of ovulation. As a result, only in females neuroendocrinologically ready to ovulate will we see sex behavior, the most important of which is lordosis. Without the swaybacked posture of lordosis, the male can not fertilize, nor can reproduction occur.

Neuroscientists recognize the idea of a "unity of the body." The brain must be producing behaviors that answer the needs of other organs in the body. Already we understand how, in the context of this unity of the body, the hypothalamus organizes sex behavior of the female to happen at an optimal time. The estrogens emanating from the ovaries tell the brain that ovulation will occur soon. Thus, the female will only expose herself to the danger of predation or the indelicacies of the male when her mating behavior would be productive. Further, the same neuropeptide that tells the pituitary to instruct the ovaries to ovulate, GnRH, also fosters lordosis behavior. The timing of ovulation and mating behavior in lower animals is carefully regulated. What about women? Physical anthropologists have claimed that even under circumstances where marital sex is more or less constant across the menstrual cycle, promiscuous sex by the woman peaks around the time of ovulation. Intriguing, yes?

So far, I've paid attention to what produces normal male sex behavior by the male, and normal female sex behavior by the female. Let's consider

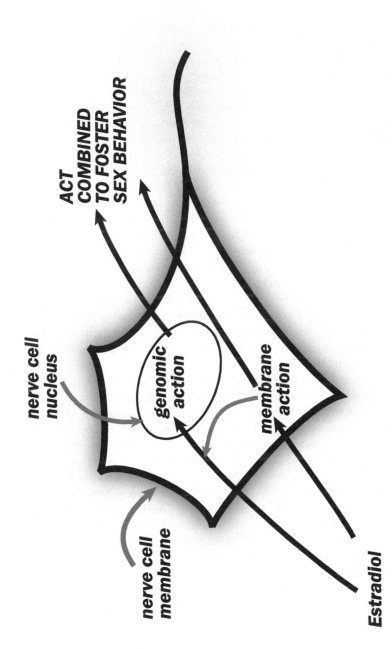

Lee-Ming Kow in my lab at Rockefeller University showed how membrane-initiated actions of estrogens in ventromedial hypothalamic neurons could combine with genomic actions to cause the female sex behavior, lordosis.

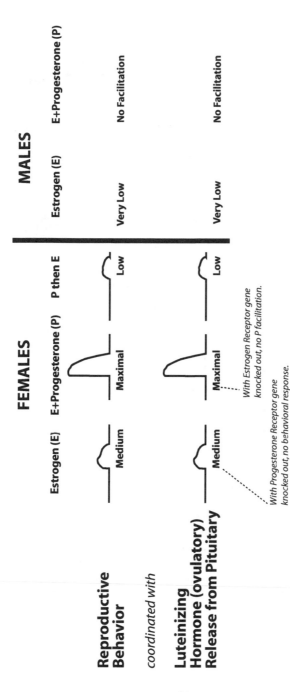

In the female brain, progesterone augments the effects of previous estrogen administration, both to produce reproductive behaviors, like lordosis behavior, and to produce the ovulatory surge of luteinizing hormone from the pituitary gland. However, if progesterone precedes the administration of estrogen, the opposite is seen. In the male brain, estrogen followed by progesterone neither produces lordosis, nor does it yield a surge of luteinizing hormone.

81

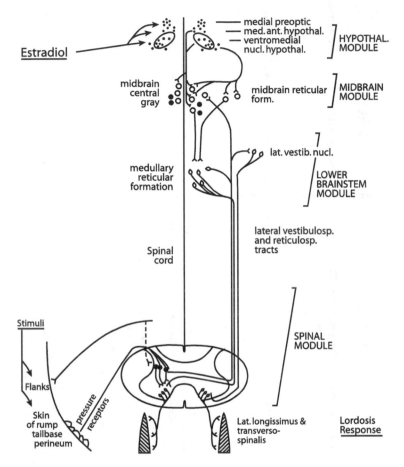

Female lab animals do lordosis behavior. Males don't. In the female, signals from the skin, indicating the male's mount, travel to the spinal cord and terminate extensively there. However, for the behavior to occur, subsequent signals must ascend from the spinal module to the brainstem module. Some sensory signals reach the midbrain. There, neurons will start a lordosis command if, and only if, estrogen-dependent signals emerge from the hypothalamus. These will have originated from prior estrogen action in the ventromedial nucleus of the hypothalamus. Once they tell the midbrain neurons what to do, the midbrain will communicate with neurons in the lower brainstem module, which in turn signal to the relevant motor neurons in the spinal cord. Those neurons activate the deep back muscles (striped), whose contractions make the lordosis posture, thus permitting fertilization by the male.

The highly receptive and aroused female laboratory animal sits in a tensed crouched posture (top). Mounting by a male causes her—via the neural circuitry shown on the facing page, and the chemical reactions shown just above—to snap into a swaybacked, convex-up posture that permits the male to fertilize.

some of the less conventional behaviors. What about times when males behave like females? It turns out that female-type sex behavior by the male is strongly inhibited by nerve cells in the forebrain that project to the preoptic area and hypothalamus. Neuroanatomists Korehito Yamanouchi and Yasumasa Arai, in Tokyo, found that that if you cut the outputs of those forebrain neurons, males will do lordosis behavior. Likewise, Kim Wallen and David Edwards at Emory University discovered that if you damage those neurons' targets in the preoptic area, males will do lordosis behavior. Conversely, if you destroy neurons in hypothalamic cell group centrally important for lordosis behavior in females, the ventromedial hypothalamus, then those females will mount, like males. Thus, harm a male behavior-producing group of neurons in the male, and get female behavior. Damage a female behavior-producing group of neurons in the female, and get male behavior.

In the human brain, the ventromedial hypothalamus, important for female behavior, and the preoptic area, essential for male behavior, are a fraction of an inch apart (see preceding figure), and in the brain of a lab animal, even less, perhaps an eighth of an inch. The two cell groups tend to have an antagonistic relation to each other. For example, in experimental animals, the same preoptic nerve cell damage that reduces male behavior actually increases female behavior by removing an inhibitory influence on female behavior.

Genetic studies of sex behaviors, by behavioral geneticist Sonoko Ogawa in my lab, revealed a surprising fact. Knocking out the gene coding for estrogen receptor-alpha wipes out female sex behavior by the female, but it also greatly reduces male sex behavior by the male. Thus, I was left with the unanticipated conclusion that an individual gene (estrogen receptor-alpha) is necessary for the normal performance of both male and female sex behaviors.

What about nonhuman primates, our most immediate ancestors? We already know that hormones coming from the ovaries, estrogens and progesterone, powerfully drive sexual desire in female monkeys. A case was made that very high levels of testosterone in females would increase the number of presentations of a courtship posture called "sexual invitations." Remember that the testosterone can be chemically converted into estradiol. The safest and biggest conclusion to date may be that in monkeys, testosterone and estradiol working together make for great sex behavior by females.

On the other hand, the strength of hormone effects on sex behavior in higher primates has been diluted by cultural and social effects. To quote Kim Wallen, Professor at Emory University: "How the physical capacity to mate became emancipated from hormonal regulation in primates is not understood. This emancipation, however, increases the importance of motivational systems and results in primate sexual behavior being strongly influenced by social context." Social context includes the nature of the male/female relationship (marital versus promiscuous), the day of the week (Sunday looks great!) and who initiates the sex.

Whether talking about laboratory animals or thinking about humans, when we talk about sexual behavior, we're talking about what the psychologist Charles Cofer called "the most powerful force motivating behavior."

Parenting Behaviors

So what exactly *are* "parenting behaviors"? According to Timothy Clutton-Brock in his book *The Evolution of Parental Care* (1991), any parenting behavior "appears likely to increase the fitness of a parent's offspring" (p. 8). These behaviors help parents to pass on their DNA. Females in a large percentage of mammalian species, including a considerable portion of human females, get ready for and do maternal behaviors naturally. These would include getting ready to give birth, breast feeding, and generally taking care of the young.

What about males? Clutton-Brock attempts to explain the circumstances under which males do and do not help females care for the young. Strategically, he guesses that males would assist under circumstances that increase survival of the young, but that advantage must be compared to the advantage of spreading his DNA around another way: by skipping the paternal care routine, so that he can use his time and energy mating with other females.

Clutton-Brock also addresses a problem of intense interest to families in which the biological father has deserted, and another man is living in the home and helping to care for the children. This is called *polyandry*. In some species, a portion of females will pair with more than one male during a single breeding season. Biological theories are hatched to try and account for this. They tend to have three themes, all of which could be right. First, if females are greatly stressed, nutritionally or otherwise,

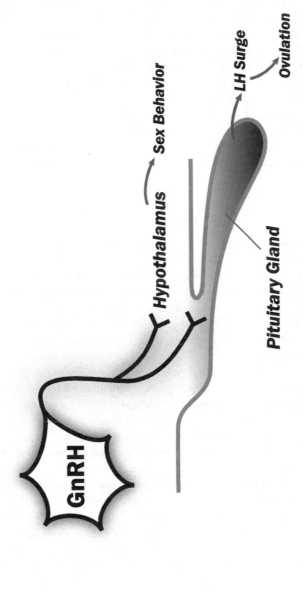

Produced in neurons in the preoptic area, the very same small neuropeptide, just ten amino acids long, that tells the pituitary gland to make a luteinizing hormone (LH) surge and cause ovulation, also acts on neurons in the hypothalamus to foster lordosis behavior. Thus, females have their mating behavior circuitry activated most highly at the time, near ovulation, when mating behavior would yield a pregnancy. This is biologically adaptive among animals. By the way, in the male, GnRH administration also elevates the amount of male-type mating behavior.

more than one male working on the problem of brood care during a single breeding season may save the babies. Second, if by polyandry the female can increase the number of babies born per breeding season, then her passage of her DNA to the next generation is enhanced. And finally, by meeting more than one male, a female may find one with a better set of genes. She will have set up a sperm competition.

In many species, strong hormone actions are much more important to trigger the onset of maternal behavior, rather than to maintain it once it has been well established. In turn, a very experienced mother—perhaps on her third pregnancy—needs much less of a hormonal "kick" than a first-time mother.

Barry Keverne, chief of the department of animal behavior at the University of Cambridge, has treated maternal behavior as the model (in German, the *bauplan*, or fundamental structural plan) for evolving the type of bond that fosters and secures all affiliative behaviors. It follows from Keverne's reasoning that, in his words, we have witnessed the "emancipation of behavior from hormonal determinants and in parallel, an increasing role for intelligent social strategies." That is, less of a brutal dominance by primitive hormonal actions, and more thinking about how best to design social roles.

Keverne and Clutton-Brock have studied animal behavior from an "ethological" point of view. To explain this point of view, I'll reflect that during the twentieth century, two major traditions in behavioral science dominated our scholarly pursuits. In the United States, animal behavior was studied with the idea of attaining the precision of a physical science, and emphasized highly controlled studies of behavior under strictly defined experimental conditions. In contrast, on the European continent and in the United Kingdom, ethology dominated. Ethology concentrates on the study of natural animal behaviors in animals' natural habitats. A past master, Sir Robert Hinde at the University of Cambridge, influenced both Keverne and Clutton-Brock who, as ethologists, came up with evolutionary arguments about why it was biologically adaptive for certain behaviors to emerge.

In comparison, Michael Numan, professor of psychology at Boston College, and Thomas Insel, a psychiatrist whose present job title is Director of the National Institute of Mental Health, are scientists who want to reduce explanations of behavior to the details of nerve cell activity—they are interested in the nuts and bolts. Their masterful book

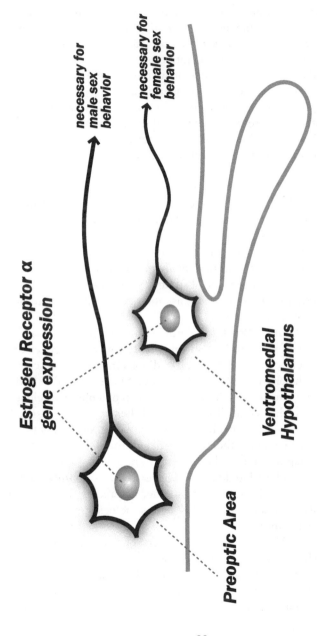

Estrogen Receptor α gene expression

necessary for male sex behavior

necessary for female sex behavior

Ventromedial Hypothalamus

Preoptic Area

This drawing looks at the basal forebrain from the left side. In front of the hypothalamus, preoptic neurons require expression of the estrogen receptor-alpha gene properly to activate male sexual behaviors. In the ventromedial hypothalamus, neurons require expression of the estrogen receptor-alpha gene properly to activate female sexual behaviors. Thus, the estrogen receptor-alpha gene is necessary for both male and female sexual behaviors.

"The neurobiology of parental behavior" gives us the modern summary of what neuroscientists know about parenting. They described how signals coming out of the preoptic area travel to the brainstem and produce the stereotyped series of complex behaviors that allow laboratory animals like mother rats and mice to take care of their young.

Numan and Insel also focused on the importance of the normal pattern of hormonal changes towards the end of pregnancy that foster maternal behavior in laboratory animals: high and rising levels of estrogens accompanied by progesterone levels that used to be high, and now are falling precipitously. That's the key. Given that pattern of changes in sex hormone level, a female mammal is much more likely to prepare for the babies' arrivals, gather them, keep them warm, protect them and nurse them. Most males simply do not have the hormonal formula—the increasing estrogen rhythm, the decline in progesterone—and thus are less likely to get the parental job done. This huge difference pertains, as well, to women and men.

Parental behavior requires marathon performances. Parents must attend constantly to their children (or pups, as the case may be) both to meet the babies' nutritional needs and to react swiftly to potential dangers threatening the babies' young lives. Susan Fahrbach, a graduate student in my lab, was interested to determine exactly where in the brain estrogenic hormones might act to foster rapid maternal behavior responses. Fahrbach neurosurgically placed tiny implants of estrogen into the preoptic area in female rats that had their ovaries surgically removed, so that they did not have any estrogens any place else throughout the body. These females now rapidly built their nests, retrieved their scattered pups, and grouped them for nursing properly, whereas before the estrogen implants they had done none of these things. Then, she went on to demonstrate that oxytocin, produced in other, different neurons in the hypothalamus, is necessary for normal maternal behavior, especially by a mother who lacks experience as a parent. Thus, another formula: in these animals, the concerted actions of estrogenic hormones and oxytocin permit rapid maternal responsiveness. Again, males do not have this formula.

The various hormonal rhythms the males don't have are involved in another type of mechanism that produces sex differences in parental behaviors, as well. Remember Gary Romano's work. Even given identical treatments of estrogenic hormones, males cannot get the estrogen-caused

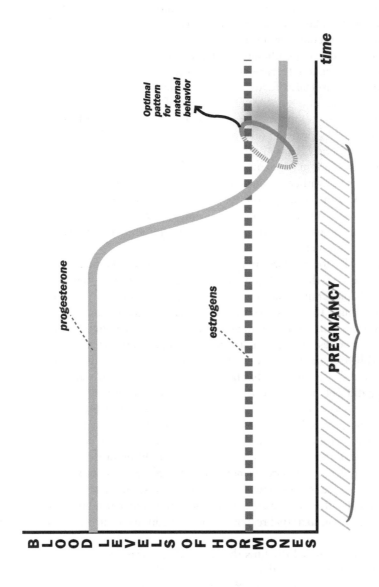

BLOOD LEVELS OF HORMONES

progesterone

estrogens

Optimal pattern for maternal behavior

PREGNANCY

time

In women and in female laboratory animals, estrogens circulate in the blood in much lower concentrations than does progesterone. To the left in this drawing, early pregnancy levels of circulating hormones are depicted. In female lab animals, at the end of pregnancy, estrogens stay at their same levels, while progesterone concentration falls precipitously. This hormonal pattern fosters the initiation of maternal behaviors: making a good nest, retrieving pups gently if they get out of the nest, nursing and licking them to keep them warm and well fed.

Successful Maternal Behavior

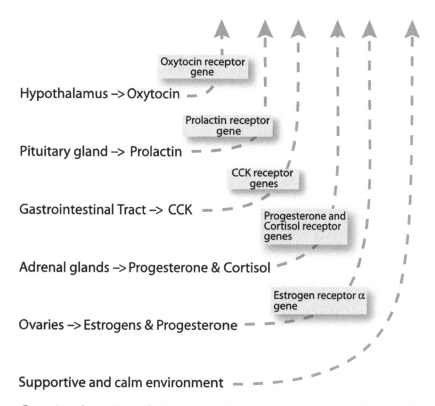

Expression of several genes for hormones and their receptors are required for successful maternal behavior. And, their actions must be supported by favorable environmental factors.

expression of the progesterone receptor gene. Nor can they increase gene expression for an opioid peptide that could help to reduce their fear and anxiety upon confrontation with strange little babies. Thus, on the genetic front, also, the males of laboratory animal species drop out of the parental care game.

Estrogens and oxytocin and genes, and that's it? No, Numan and Insel make it clear that even for simple laboratory animals, females with previous experience raising babies are faster and more efficient than those without experience. It's not a matter of "maternal" versus "not maternal," but instead, a matter of speed and quality of care. For example, among nonhuman primates, mothers who have been through child raising before are less likely to reject, neglect, or abandon babies than those who are trying to nurture a baby for the first time. Reading between the lines, it looks to me as though anxiety associated with certain environments, and the unfamiliarity of the baby itself, have something to do with the most troubled mothers. Think about this with respect to what we read in the newspapers—mothers, especially young women in difficult situations. Numan's and Insel's coverage of the scientific literature suggests to me that a pharmacological approach is not necessarily required in such situations, but instead a psychotherapeutic and educational approach. And the more demanding the circumstances impeding the mother's efforts, the more important the education and support. Put this all together and you have a complex series of caring behaviors regulated by several hormones and dependent on experience.

Finally, we also have to deal with opportunities offered by, and the problems caused by, the father. On the one hand, some species of mammals including humans have the possibility of both fathers and mothers caring for the young. Testosterone, chemically converted to estradiol, as I illustrated earlier, can help foster paternal behavior. On the other hand, the parental instincts of laboratory animals are much greater for the females than the males. If the environment or the strangeness and unfamiliarity of the babies upset the male, he is going to attack. That's still another massive example of sex differences in parental behavior.

The Story So Far

At this point in our story, all the basics have been covered. We learned what makes a biological male and a biological female (X and Y sex chromosomes). We've discussed gene expression and differences in male and female brains. In this chapter, we've begun to explore hormones and their implications for mating and parenting behaviors. Next, we will look at males' aggression.

SIX

Males Fighting

What do aggressive behaviors do for us? In the words of Carolyn and Bob Blanchard, renowned ethologists at the University of Hawaii, three things: they control another individual, they assert authority, and they extract revenge. For humans, male aggression and violence are huge problems for societies across the planet. And these aggressive acts range from the symbolic challenges of teenage boys by teenage boys, to violent criminality by males of various ages.

The sex difference starts early in life, and throughout life can be observed in primates other than humans. Young male monkeys, for example, exhibit a lot more rough and tumble play than young female monkeys. Within American families, more than 90% of violent acts are initiated by an adult male. Murders of males by unrelated males in American society, as well as other societies, follow a lifetime curve that is closely paralleled by the curve of testosterone concentrations in the bloodstream. Altogether, it looks as though androgenic hormones like testosterone, acting on the male brain, predispose some males in some situations to commit violent acts.

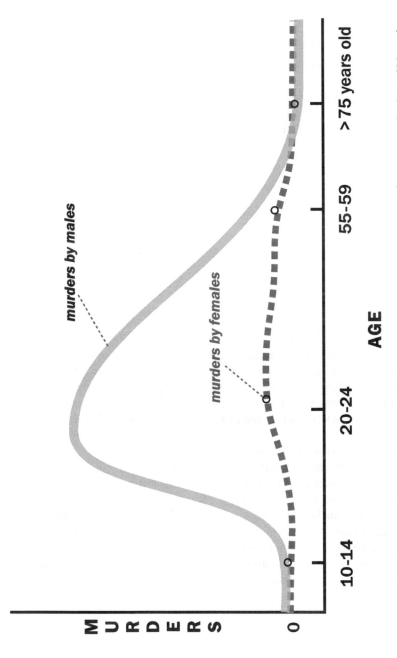

MURDERS

AGE

10-14 20-24 55-59 >75 years old

murders by males

murders by females

Murders of males by unrelated males occur most frequently during the age span when males' testosterone levels are highest. As summarized by social psychologists Margo Daly and Martin Wilson, at McMaster University in Canada, this is an international, cross-cultural pattern. Most murders are committed by males, although now there are more murders by females than there used to be.

Yes, the big story with male–male fighting is testosterone. This hormone, secreted from both the testes and the adrenal gland, is a steroid, a flat, rigid molecule, fairly simple as biochemicals go. But its mechanisms of action in the brain to produce aggressive behaviors are far from simple. First, in addition to testosterone hitting neurons head on, this hormone can "go through a sex change" and act a second way. A series of three chemical reactions change testosterone into estradiol, and in just the right places on the brain, estradiol can foster male–male aggression as well. Second, not only does testosterone drive aggression in adulthood; it also has major effects in the developing brain. The behavioral endocrinologist David Edwards, now at Emory University, was the first to show that early exposure to testosterone, during the period for sexual differentiation of the brain, virtually guarantees that experimental mice, having reached adulthood, would show high levels of male-type fighting behavior even if, genetically, they were females.

For understanding hormone effects on aggressive behavior by men, consider testosterone loss versus testosterone overload. With respect to testosterone loss, obviously, eunuchs were never known for exhibiting much physical aggression. Studying the opposite side of the testosterone situation, high testosterone, Harrison Pope, at the Harvard Medical School, was the first medical doctor to provide convincing evidence for strong effects of androgenic hormones on human aggression. He described a range of changes in mood and behavior in a groundbreaking series of scientific papers. Prominent among these changes was the androgenic hormonal facilitation of aggression, ranging widely from feelings of hostility to homicide.

For our teenagers, the story with synthetic androgenic hormones— anabolic androgenic steroids (AAS)—is much more grim. First, of course, when they foster aggression they throw a boy into a social role that perpetuates hostile acts and causes others to expect hostile acts from that teenager. Further, their use has escalated, in some cases to enhance athletic performance and in other cases to "look good"; that is, to have bigger muscles, thus improving a guy's body image. Used in this way with high doses, they can cause "roid rage." Worse, their potential for long-term effects on aggression is unknown.

As I wrote about in my book, *The Neuroscience of Fair Play*, Marilyn McGinnis, a prominent behavioral endocrinologist now at the University of Texas at San Antonio, has investigated their behavioral

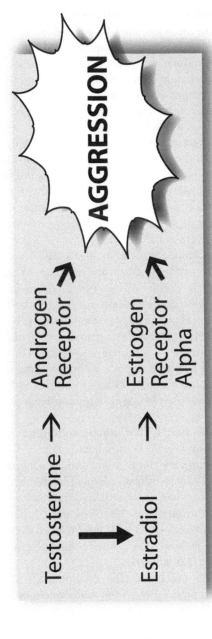

Acting in the male brain, testosterone can foster a male's aggression by acting in its own chemical form and binding to the androgen receptor, and also by being converted enzymatically in the brain into estradiol, thence to bind to estrogen receptor-alpha.

effects in laboratory animals. Comparing several steroids under different experimental conditions, she found that their effects on aggressive behavior depend upon exactly which AAS is being used (some increase aggression while others actually decrease aggression), but also depend upon the provocation by environmental cues. Hormone-loaded males were more aggressive toward other hormone-loaded males, were more aggressive when defending their home cage, and were more aggressive in response to pain. McGinnis feels that androgenic hormones do not simply drive aggression pure and simple. Instead, in her words, "AAS's sensitize animals to their surroundings and lower the threshold to respond to provocation with aggression." Since she studied pubertal male animals, her results may be relevant to the troublesome incidence of violent acts among adolescent boys and young men. As for long-term implications of AAS used in the teenage years, I note that some researchers associate high testosterone levels with the development of antisocial personality characteristics.

Connections of high testosterone levels with aggressive behaviors are not limited to our American culture. In San Sebastian, in the Basque country, and in Seville, Spain, endocrinologist Aitziber Asurmendi and his colleagues found a highly significant correlation between androgenic hormones in the blood and the tendency toward provocative behaviors, even in 5-year-old boys. These boys would invite a fight. And an international team led by endocrinologist Irene van Bokhoven in the Netherlands studied adolescent boys and saw similar correlations between testosterone and self-reported delinquent behavior, and proactive and reactive aggression. At age 16, according to their article in a recent issue of the journal, *Hormones and Behavior,* Dutch boys who had already developed a criminal record had higher blood testosterone levels.

What are the cellular mechanisms by which testosterone works to affect aggression? Some of the routes are indirect, and do not involve the brain at all. Testosterone and other androgenic steroids make muscle cells grow. Those large muscles would make a guy more able to fight victoriously. They also would make him more confident that he could win, and thus encourage him to start a fight. Put two of these guys together and you have trouble.

In the brain, as well as in the muscles, testosterone works by entering the cell and binding to a large protein called the *androgen receptor*. Once bound to testosterone, this protein enters the nerve cell nucleus and

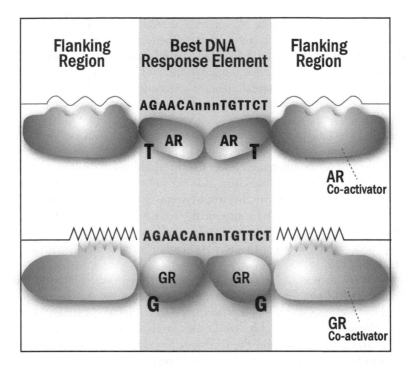

Having bound testosterone (big T), the androgen receptor (AR) is a large protein that has a favorite sequence of DNA nucleotide bases upon which to fasten (DNA bases = A, adenine; G, guanine; T, thymidine; C, cytosine. n = any base; it is a spacer) Notice that a glucocorticoid hormone receptor (GR), having bound a glucocorticoid hormone such as cortisol, also preferentially fastens onto the same DNA sequence. Thus the question: Where do the specificity and particularity of testosterone and cortisol actions come from? Diane Robbins, professor of genetics at the University of Michigan, discovered that the specificity of hormone action in this case derives from specialized proteins that join the hormone receptors in the controlling regions of hormone-sensitive genes, proteins called *co-activators*, and that these co-activators recognize regions of DNA that flank the primary DNA response element. These flanking regions are different in androgen-sensitive genes than in cortisol-sensitive genes.

attaches itself to specific portions of DNA upstream of testosterone-sensitive genes. Thus attached to DNA, the androgen receptor increases or decreases their expression, depending on which gene we are talking about. This train of mechanisms is slow; it takes hours. A second way by which testosterone affects nerve cells is fast, similar to mechanisms I talked about before. John Nyby, a biologist at Lehigh University, and others, have reported evidence that testosterone, like estradiol, can work at the membrane of nerve cells thus to affect the activity of those nerve cells within seconds or minutes, causing a rapid change in behavior.

Genes Count

Knowing that males initiate a large preponderance of violent acts, I want to focus on the Y chromosome first. Robin Lovell-Badge at the National Institute for Medical Research in London, as noted in Chapter 2, discovered the SRY gene on the Y chromosome that sets off the cascade leading to a masculine type of sex differentiation, including the development of the testes, the endocrine organs that will produce testosterone in the first place. But even before Robin Lovell-Badge's work, Steven Maxson and his colleagues at the University of Connecticut had analyzed the Y chromosomes of male mice and found that, quite separate from SRY, a region of the Y that harbors genes not on other chromosomes is especially important for fostering aggressive behavior. These Y-chromosome effects on aggression give us some of the simplest examples of how genes can influence behavior.

Thus, Y chromosome genes are involved in the regulation of aggression in males. Further, genes are often brought into the behavioral picture by showing inheritance of a behavioral trait. We clearly know that tendencies toward male aggressive behavior can be inherited because of animal breeders' abilities to produce animal strains, lines of animals that have been purposively bred to produce different levels of combativeness. Fish, dogs, birds, rats, mice and horses, all would give examples of this. And, human twin studies have shown a strongly inherited tendency toward aggression, accounting for perhaps 50–60% of the variation among children in their most disruptive, obnoxious behaviors.

Admittedly, these are early days in the functional genomics of aggression and violence. Nevertheless, I can give specific examples of genes influencing aggression. I've already mentioned the genetic contributions from the Y chromosome which tell us to expect that males will, on the average, be more aggressive than females. After that simple statement, however, things get more complicated. A newspaper writer might, for example, talk about "a gene for aggression," but from the panoply of genes

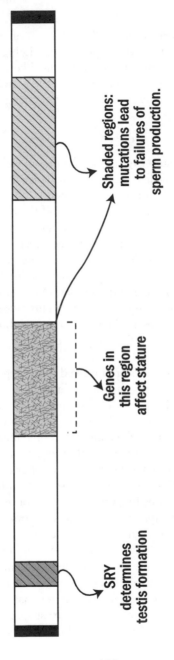

Shaded regions: mutations lead to failures of sperm production.

Genes in this region affect stature

SRY determines testis formation

Only parts of the tiny Y chromosome are unique to the Y (when comparing to the X chromosome), and three of them are drawn here. Notice the small region at one end of the Y that houses the critically important SRY gene.

involved that I will illustrate now, we know that normal, biologically regulated aggression (as opposed to psychopathic violence) will depend on patterns of gene expression orchestrated over time, not on just a single gene.

Talking, therefore, about testosterone leading to aggressive behaviors, I am not portraying a simple, single-factor picture. There are at least three levels of complexity. First, remember that testosterone can increase aggressive behavior not only acting as its own, unaltered chemical self, but also after chemical conversion to estrogen. Second, lots of genes are involved. And third, I mentioned those rapid actions of testosterone studied, for example, by John Nyby. And now the story gets even more complex.

If Tom Cruise had not accepted the title role in the movie "The Last Samurai," Sonoko Ogawa in my lab could easily have played the part out of her normal personality. A scientist for all seasons, she attacked problems of the relations between genes and aggressive behaviors with an intellectual ferocity and an emotional determination beyond compare. One of the subjects that grabbed her attention was this. The modern steroid hormones that we know and love represent a flowering, during evolutionary time, of many different types of hormone receptors derived from a simple, single, ancient receptor. Along the way there was a primitive estrogen receptor that morphed into two related types: estrogen receptor-alpha (ERα) and estrogen receptor-beta (ERβ). These are gene duplication products, two different estrogen receptors that display overlapping but different properties. For example, estrogen receptor-alpha is required for female sex behavior, whereas estrogen receptor-beta is not.

To the point, Sonoko discovered that knocking out estrogen receptor-alpha in female mice greatly *increases* aggressive behavior, whereas knocking out estrogen receptor-beta *decreases* it. In fact, Sonoko and I found that knocking out estrogen receptor-alpha causes females to behave like males and to be treated by other mice as males. Even more surprising, the pattern of results is exactly the opposite in males. In males, knocking out estrogen receptor-alpha abolishes aggressive behavior, whereas knocking out estrogen receptor-beta increases it. Thus, deletion of the gene duplication products can have opposite effects on aggression and, moreover, the pattern of results is opposite between males and females. The effect of a gene on behavior can depend on the gender in which it is expressed.

Knocking out gene →	Causes:	In Males	In Females
for Estrogen Receptor α	Aggression	⬇	⬆
for Estrogen Receptor β	Aggression	⬆	⬇

In evolutionary history, there once was only one gene coding for an estrogen receptor. Then, that gene type is thought to have duplicated, producing estrogen receptor-alpha and estrogen receptor-beta, similar to each other but not identical. Surprisingly, Sonoko Ogawa in my lab at Rockefeller University found out, with respect to estrogen actions in the brain, that knocking out the gene for estrogen receptor-alpha has the opposite effect on aggressive behavior as knocking out estrogen receptor-beta. Even more surprising, for either of these two genes, knocking it out has opposite effects on aggression by females compared to aggression by males.

Later, I will talk about the involvement of neurotransmitter serotonin in the regulation of aggression. Studies of human genetics have, as well, pointed to serotonin. Klaus Peter Lesch, at the University of Wurzberg, Germany, has done extensive genetic work to investigate which genes involved with serotonergic transmission might regulate aggression. In Lesch's studies, not only several serotonin receptor genes, but also genes coding for a serotonin transporter—a protein that picks up serotonin out of the synaptic cleft, and puts it back in the nerve cell it came from—and a serotonin-breakdown enzyme were associated with altered impulsivity and aggression in human volunteers.

Gene Products, Neurochemicals Directly Involved

One of the oldest neurotransmitters in the brain, the excitatory neurotransmitter glutamate, is clearly involved in the hypothalamic control of male aggression. In a part of the hypothalamus whose electrical stimulation rapidly evokes attack behavior, Eva Hrabovszky and her neurobiological colleagues in Budapest found an overwhelming preponderance of glutamatergic neurons. Then, neuroanatomist Allan Siegel, of Rutgers University, showed that these hypothalamic neurons with glutamate send axons to the midbrain central grey, acting there through specialized glutamate receptors on midbrain neurons to cause vicious attacks. Interestingly, a single enzyme converts this ancient, excitatory transmitter into GABA, an equally ancient inhibitory transmitter. Leslie Henderson and Ann Clark, physiologists at Dartmouth Medical School have been able to implicate GABA neural transmission in producing androgenic effects on behavior, but exactly how the GABA effects are played off against the excitatory glutamergic effects in the male hypothalamus remains to be discovered.

As introduced earlier, a huge neurochemical subject for the discussion of aggression is the transmitter serotonin—its synthesis, its receptors and its breakdown. Berend Olivier, a pharmacologist at the University of Utrecht in the Netherlands, has charted a large number of examples showing that drugs increasing serotonergic function will reduce aggression by males. Stephen Manuck and his neurochemical colleagues, writing in a book edited by the authority Randy Nelson, described their evidence that low levels of serotonin in the brains of male rhesus monkeys caused the animals "to have fewer social companions and spend less time in passive affiliation or in bouts of grooming with conspecifics." Importantly, high serotonin in the brain was associated with low levels of aggression, whereas low levels of serotonergic function in the brain caused a high propensity to initiate high-intensity aggressive behaviors such as chasing and physical assaults. They also reviewed the literature on the relations between serotonin and aggression in humans. Clearly, low levels of serotonergic function in the human brain predispose to high levels of aggression by men. For example, Emil Coccaro, a psychiatrist at the University of Chicago, has tied low levels of serotonergic function to extreme forms of impulsive aggression that reflect long-lasting tendencies in certain patients with personality disorders. Mutations in the gene for the enzyme that manufactures serotonin are associated with high levels of impulsive aggression. Some of the strongest correlations come from cases of homicide in which the murderer killed a sexual partner. That's the case for outwardly-directed aggression. Dr. Coccaro also considers "inwardly directed aggression," manifest in its most extreme form, suicide. Diminished serotonergic activity in the human brain predicts an increased risk for suicide. Manuck et al. are concerned to explore a key concept in the neuroscience of aggression, *impulsivity*. As opposed to slowly developing, calculated aggression, impulsive aggression bursts out suddenly and is accompanied by emotional expressions such as anger and rage. Manuck takes the view that serotonergic effects on aggression do involve these emotions, and that drugs reducing impulsive aggression by boys and men are likely to work, in part, because they also reduce negative emotions in general.

Another way of linking serotonergic function with reduced aggression is to think about the gene coding for the enzyme called MAO-A, which is responsible for breaking down the serotonin molecule. Avshalom Caspi, Terrie Moffitt, and their colleagues in the Institute of Psychiatry at King's College, London, concentrated on this X-chromosome gene because previous research with mice and with men indicated that its absence caused antisocial behaviors. Caspi et al. asked a different kind of question. They wanted to know how different levels of expression from the MAO-A gene in young men might affect the damaging consequences of having been maltreated as children. A short form of the MAO-A gene produces an MAO-A enzyme that has abnormally low enzymatic activity. It looks as though the form of this gene that makes an inefficient enzyme—a form that occurs in about 34% of individuals—predisposes a person toward high levels of impulsive aggression. Young men with this MAO-A genetic allele had higher levels of conduct disorder, greater dispositions toward violence, higher frequencies of antisocial personality disorder, and more convictions for violent criminal offenses. Most important, these convincing results only appeared among young men who had been severely maltreated as children. Two forces for violence, early maltreatment and a genetic alteration, multiplied each others' effects on antisocial behavior. Thus, in this extremely important domain having to do with the genesis of violent behaviors by young men, Caspi et al. brilliantly revealed an example of significant gene–environment interaction. But serotonin neurobiology by itself is not sufficient to regulate aggression. Enter the peptide, vasopressin.

Vasopressin (VP) is a tiny bit of a protein, only 9 amino acids long. Several studies by well-known professors of psychology—Craig Ferris at the University of Massachusetts and Elliot Albers at Georgia State University—have connected vasopressin with aggressive behaviors in males. Microinjecting vasopressin into certain parts of the hypothalamus will facilitate aggressive behaviors. Microinjecting a chemical that blocks

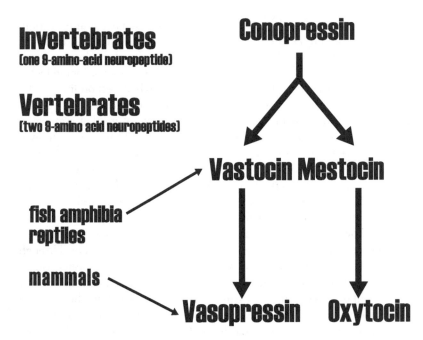

Invertebrates
(one 9-amino-acid neuropeptide)

Vertebrates
(two 9-amino acid neuropeptides)

Conopressin

Vastocin Mestocin

fish amphibia reptiles

mammals

Vasopressin **Oxytocin**

Simplified sketch of how neuropeptides, characterized by having nine amino acids and a special bent shape, and all concerned with body fluid distribution, evolved into the two—vasopressin and oxytocin—found in our own bodies. I enjoy describing their roles in regulating social behaviors in this chapter and the next.

access to VP receptors will reduce them. The simplest summary of those kinds of data, right now, would say that vasopressin promotes aggression in males. Elliott Albers and his Georgia State colleague Kim Huhmann feel that one of the ways various kinds of social manipulations that increase aggression in the laboratory might work is by increasing the amount of vasopressin binding to its receptors in the hypothalamus.

Vasopressin evolved from an older gene that coded for a 9 amino acid neuropeptide. That older gene also yielded vasopressin's gene duplication product, oxytocin. Oxytocin promotes friendly behavior by females, as I'll talk about in the next chapter. Thus, vasopressin for aggression by males, versus oxytocin for friendly behaviors by females.

New results link these findings to human social behavior. Physiologists R.R. Thompson and J. Benson, working at Bowdoin College and Harvard

Medical School, looked at how VP influences social communication. Men administered VP displayed more *un*friendly facial motor patterns, as measured by electrical activity of certain facial muscles, when responding to pictures of the faces of unfamiliar men. But women affected by VP displayed more friendly facial motor patterns when responding to the faces of unfamiliar women. These results with men fit in very well with the literature on animal brains and behavior, a literature that connected VP and male—76 male aggression, but the sex differences in human social behavioral responses were unexpected by the authors.

Then came a surprise: a gas in the brain! Nitric oxide is not a regular neurotransmitter. It is actually a gas. The gene coding for an enzyme that makes nitric oxide in the nervous system, neural-nitric oxide synthase (nNOS), plays an important role in the regulation of aggression. Randy Nelson, now at Ohio State University, discovered by accident that male mice lacking this gene were highly aggressive toward each other. Through a long and systematic series of studies, Randy and his students used a variety of genetic and pharmacologic tricks to prove that nitric oxide must be thought of as reducing aggression. That is why abolishing the function of nitric oxide synthase increases it. Another fascinating finding from the Nelson lab: female mice lacking the nNOS gene never showed unusual aggression. Therefore, these scientists found an example in which the effect of a gene on aggression depends on the sex of the animal in which that gene is expressed. Nelson and his team also produced a nice illustration of an interaction between gene and environment. They could significantly reduce aggression in the nNOS gene knockout males if they are housed together all the time from weaning until testing, compared to the level of aggression of males housed in isolation. The social environment in which the animal is raised determines whether or not there will be a genetic effect on behavior.

Most neuroscientists, when they are studying brain mechanisms that regulate aggression, think of the fighting itself. It turns out, however, that a deeper and more primitive force, almost a "dark energy" of the brain, contributes to the production of aggressive acts in an essential manner.

"Dark Energy" of the Brain Contributing to Aggression

This "dark energy" is called *generalized brain arousal*, or generalized CNS (central nervous system) arousal. The examples I've presented of genes for nuclear receptors, neurotransmitters, and neuropeptides participating in the regulation of aggressive behaviors, represent just the beginning for my field of scientific work on aggression and other instinctive behaviors. Complexities abound. First, as I've said before in my book, *Brain Arousal*, we must consider that an important factor in allowing a mouse or a man to be aggressive is his level of generalized CNS arousal. This primitive, powerful force for arousal underlies all emotional expressions, and provides the emotional "fuel," the psychic power for aggression. I have documented elsewhere that more than 100 genes are involved in the control of arousal. Many of these certainly will affect aggression. Secondly, in order to keep this section of the book readable, I have treated males' aggressive behaviors as though these were all similar to each other. But, really, different kinds of aggressive behaviors—territorial attacks, predatory aggression, emotional explosions, defensive behaviors, verbal aggression, and so forth—are not identical to each other, and are likely to bring in the influences of different genes we have not yet considered.

These are exciting times in research on aggression. Indeed, geneticist Edward Brodkin at the University of Pennsylvania Medical School has recently used a sophisticated genetic technique to show that as-yet-unidentified genes will be discovered as affecting aggression in male mice. The genetic and neurochemical analyses of different types of male aggression and violence are just beginning. But they all depend on generalized brain arousal.

Continuing on the importance of generalized brain arousal, I return to serotonin. Working with serotonergic neuronal systems, the Dutch neuropharmacologist Berendt Olivier has developed drugs that would reduce aggressive behavior. Indeed, he found some whose properties to influence behavior were such that he called them *serenics*—the male animals given these drugs seemed "serene." I noted, however, as Berendt was speaking at a recent meeting, that these drugs also nonspecifically reduced the overall likelihood that the animals would emit voluntary motor activity. Thinking about how I claimed in *Brain Arousal* that high levels of central nervous system arousal are necessary for aggressive behavior, I realize that the side effects of Berendt's serenics represent one of the important mechanisms by which his drugs work on aggression. I now must speculate, therefore, that Berendt's drugs are reducing aggression purely because they reduce generalized brain arousal. This thinking lays bare a major problem in the field: how do we reduce aggression without reducing the generalized CNS arousal that powers the expression of all of our emotions?

Where in the Brain Does the Fighting Come From?

Let's start at the front of the brain. A small region in the middle of the fore-brain called the *septum* is very important for the regulation of aggression. It powerfully inhibits aggressive behavior. If you damage the septum in an experimental animal you produce a syndrome called "septal rage." Caroline and Robert Blanchard point out that the animal's enraged attack may actually derive from underlying fear or panic. Nevertheless, if you stick a pencil into the cage of a normal male rat, he may sniff it. But if you stick it into the cage of a rat whose septum has been destroyed, he viciously will attack and bite it. I had grown used to handling septally damaged rats and, unwisely, just hours before a lecture trip to Budapest, I relaxed during handling of the very last animal. He bit me so badly that I traveled with a very badly swollen hand, and had to admit the problem during my lecture in Budapest.

On the other hand, another cell group in the forebrain, the amygdala, which I discussed in Chapter 2, is also important for regulating aggression, but in a bidirectional way, up and down. Although popular summaries of brain science recite a single name *amygdala*, that name really refers to a tightly packed bunch of more than ten nerve cell groups. About aggression, the very briefest summary states that activity in one part of the amygdala, the medial amygdala, stimulates aggressive attacks, and that activity in a different part of the amygdala, the central amygdala, suppresses attacks in males. In fact, abnormal activity of the medial amygdala can be associated with psychopathologic aggression in patients. And that psychopathologic aggression can be reduced by surgical removal of the medial amygdala. How might the amygdala function with respect to aggression? Well, there is a wrong way to think about it and a right way. The amygdala's outputs do *not* cause these alterations in emotional and social behavior in isolation, by simply acting as an autonomous command center for aggression. Instead, the best way is to think of the amygdala as working on aggression by modulating behavioral responses to a wide variety of provocative, emotionally laden stimuli. Outputs from the amygdala then determine whether the responses to such provocative stimuli will be aggressive, or not.

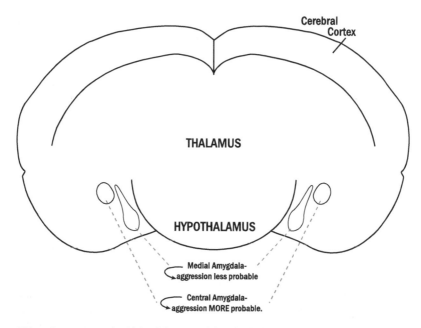

Although many people think of the amygdala as a single structure on the left and right sides of the basal forebrain, it actually comprises more than ten discrete nerve cell subgroups. Different subgroups have different effects on aggressive behaviors.

These forebrain structures important for regulating aggression send some of their axons to a small, primitive part of the midbrain called the *central grey*. That connection is important, because electrical stimulation in the midbrain central grey causes rage-like responses in animals. This may be, in part, because neurons in the central grey also signal pain.

Since pain, uncomfortable temperatures, social stresses and other unpleasant environmental stimuli represent—in the words of social psychologist Leonard Berkowitz, speaking from the University of Wisconsin—"environmental stimulation to affective aggression," I must conclude that sensory pathways carrying signals for stress and signals of pain are involved in the neural regulation of aggression, even though those pathways are not *exclusively* involved with aggression.

Among these various regions of the brain, where do the androgenic hormones, discussed previously, impact nerve cells to promote aggressive behavior? For sure, these hormones act in the septum, the amygdala, and its closely related structures. An authority like neuroanatomist Geert DeVries, at the University of Massachusetts, would tell us that andro- genic steroid hormones promote aggressive behaviors most powerfully by acting on neurons in a cell group that rides herd on some of the outputs of the amygdala, in a tongue twister called the *bed nucleus of the stria terminalis* amygdala. Importantly, DeVries has documented significant sex differences in VP in these areas of the forebrain—with, as you would expect, males having more VP cells and denser VP projections than females. To put it another way, structural differences in this part of the forebrain are due primarily to sex differences in VP cells in the amygdale, and in the bed nucleus of the stria terminalis. For you to pic- ture this, the bed nucleus of the stria terminalis is a sinuous group of nerve cells in the forebrain that help relate amygdala activity to the hypothalamus and the rest of the brain. Its expression of the androgen receptor is much greater in males than in females, and its expression of androgen receptor is testosterone-dependent, according to the data of endocrinologist Neal Simon at Lehigh University. Put all these hormone effects together and you have a rush of testosterone-dependent signals coming out of the forebrain and heading for the hypothalamus and the midbrain.

Thus, following the thoughts of DeVries, I would put my money on the steroid-sensitive vasopressin neurons, in the amygdala and bed nucleus of the stria terminalis, for mediating the biggest effects of testos- terone on aggression. Practically all of these neurons have androgen receptors. Not only that—vasopressin binding in the parts of the hypo- thalamus where vasopression stimulates aggression is itself testosterone dependent. In turn, androgenic hormones stimulate expression of the vasopressin gene; and vasopressin robustly stimulates aggressive behavior in a wide range of species. Therefore, this vasopressin route is one causal route by which androgens foster aggressive behaviors in males.

Returning to the septum, I have to emphasize the importance of androgenic hormone effects on the GABAergic neurons there. Those neurons are chock full with androgen receptors. Importantly, these GABA neurons are the most conspicuous targets of the steroid-sensitive vaso- pressin neurons in the bed nucleus of the stria terminalis. Injections of vasopressin into the septum, which activates these GABA neurons, can actually stimulate aggressive behavior. By stimulating release of GABA, an inhibitory neurochemical, vasopressin will reduce activity in a brain region that inhibits aggression, thus releasing aggressive acts by males.

Subtleties

Some causes of male aggression in animals and humans are subtler than straightforward levels of androgenic hormones in the blood. First, the brains of some males will be more sensitive to a given amount of testosterone than others. That could be due to androgen receptor levels but there a lot of other neurochemical factors as well (serotonin levels, serotonin receptors, proteins that cooperate in the cell nucleus with the androgen receptor) that account for individual differences in sensitivity to testosterone.

Second, we must always consider interactions between hormones and genes, on the one hand, and the environment, on the other. In humans, the most obvious and important aspects of the environment have to do with stimuli coming from the person with whom a male might or might not fight. As mentioned by Professor Marilyn McGinnis at the University of Texas, testosterone doesn't just push a guy over the edge, into a brawl. Instead, testosterone makes the guy more sensitive to provocation. If a male does not get that provocation, or if he is simply not receiving the stimuli that might have provoked him, then he'll not turn mean.

Finally, I emphasized generalized brain arousal as providing the psychic force that permits aggressive acts. Realize, now, that another aspect of the environment that interacts with genetic influences on aggression derives from how much that environmental situation has stimulated generalized brain arousal. Brain arousal can influence human aggression in at least two ways, according to psychologists Craig Anderson and Brad Bushman at the University of Iowa. First, brain arousal supplies the neural energy that will direct any vigorous behavior, especially one that has such strongly emotional content as aggression. Second, brain arousal is so global, according to my own theory portrayed in *Brain Arousal*, that it can "bleed over" from other sources, and then can be mislabeled as 'anger' in the social situation at hand. One such source is stress— it requires arousal. Alan Leshner, currently chief executive officer of the American Association for the Advancement of Science, has documented that sudden increases in stress hormones fuel aggressive acts.

And reduction of brain arousal by tranquilizers inhibits aggression. If a man's aggressive act were considered as a physical vector, the angle of that vector might represent the nature and target of his aggression, but the length of the vector, the explosive force of his angry act, depends on generalized brain arousal.

Multiple Causes in Genes and Environment

So, now we know well that part of a male's tendency to be aggressive is inherited, but that the situation is far from simple. Consider all of the chemicals, the gene products I have talked about, that mediate effects of testosterone on aggression: hormone receptors, neurotransmitters, neuropeptides, enzymes and even a gaseous transmitter called nitric oxide. That means that male aggression must be a "multigenic" trait and, therefore, that small modifications in any of a large number of genes may modify a man's social behavior.

A young man's inclination toward violent acts cannot be considered independent of his environment. If he is provoked, there may be a fight. Further, environmental stimuli that, in a nonspecific manner, raise his level of generalized brain arousal will increase his tendency to respond to provoking stimuli. That is part of the gene/environment interaction.

How much can society do to reduce provocation and support an adolescent boy to develop toward his normal, crime-free introduction to society? As it turns out, a lot. James Gilligan, former Harvard faculty and head of psychiatric services for the Massachusetts prison system, worked with John Devine in New York City to conceive of a public health approach to violence by adolescent boys. As portrayed in their Annals of the New York Academy of Sciences, their program includes some very tall orders. Among the things to do are to use massive social programs to reduce extreme socioeconomic disparities and consequently avoid humiliating the young boy. Also, encouraging smaller school sizes, thus to avoid anonymity; promoting rituals that offer positive visions of the boy's adult roles in society, and moving fast to reduce consequences of impulsive behavior by the adolescent boy. Given the boy's testosterone levels and his male-differentiated brain, Gilligan and Devine want a wide variety of environmental influences to mitigate social and economic influences that might have led the boy down an antisocial, aggressive path of development. Thus, on top of all the biology about sex differences in aggressive behavior that I have covered here, I realize that layers upon layers of psychological and social factors also play their parts.

Mothers Defending Nests

Although I have spent much time with the major story, male aggression, I should at least mention a situation in which females can be depended upon to show aggressive behavior.

Among animals, males fight for females and females fight for their babies. As David Edwards, at Emory University, has nicely put it, "Maternal aggression in (laboratory) mice and rats resembles internal aggression in form, but the hormones responsible for its expression are secreted by the ovary." For female laboratory animals, a clear but somewhat complex pattern of hormonal change is necessary for the best maternal behavior. Estrogens, high throughout pregnancy, should remain high for the mother to be very nurturing toward her pups after they are born. But progesterone levels in the mother's blood, high during pregnancy, must plummet if the female is going to take adequate care of the babies: that is, if she is going to make a nice nest to keep them warm, to retrieve them to the nest if they get out, and to nurse them. Now, what happens if a strange male approaches and threatens the nest? If, and only if, those progesterone levels have remained very low, the mother can defend the nest fiercely. One of the strongest findings with respect to female aggression under a variety of conditions is that progesterone inhibits aggressive behaviors. For example, aggression against the intruder male is sluggish, late in its appearance and not very intense if it does indeed appear. But, in the absence of high levels of progesterone, females will fight, indeed, to protect their homes and their babies.

The Story So Far

There are plenty of reasons for women to get as angry as men, and to use sophisticated modern expressions of aggression as a consequence. Here I have burrowed into the detailed neuroanatomical and neurochemical mechanisms by which primitive forms of aggression, shown by lower animals, come about. While females defending their young can also become quite ferocious, brain mechanisms for their prosocial, friendly behaviors seem often to predominate. I'll talk about these mechanisms in the next chapter.

SEVEN

Females Befriending (Males, Too)

Neuroscientists are very interested in discovering mechanisms in the brain that lead to our social behaviors. As we study social behaviors in laboratory animals, certain sex differences become obvious. Let's start with the first step, social recognition.

In the movie, "The Changeling," the mother Christine (Angelina Jolie) knows without any doubt that the child returned to her by the Los Angeles police is not her lost son. Mothers know. Same for animals. The Harvard anthropologist Lucien Barbash-Taylor has filmed examples of female sheep, mothers who are being asked, in the name of farming efficiency, to take care of newly born lambs that are not their own. But do they recognize this? They can be fooled. If, and only if, within 24 hours of their giving birth, the foster lambs are covered with the foster mother's own amniotic fluid, the farmer has a win. That ewe will "recognize the kid as her own" and will take care of the foster lamb. This maternal behavior gives us an example of social recognition, and social recognition lies at the basis of all loving care. In fact, as I'll describe in this chapter, social recognition lies beneath all friendly behaviors. To befriend someone we must first recognize him and know that he will not harm us.

And Who Might You Be?

The people we grow up recognizing, and usually have a good, warm feeling about, and from whom we certainly should not expect harm, are those in our own family. Historically, we had the concept of an economic unit in which the mother, father and children work effectively together; for example, on a family farm. The physical anthropologist Helen Fisher tells us in her book *Why We Love* that this kind of bonding became especially important for women after humans developed a striding walk. That is to say, this kind of pair bonding must have evolved by at least three million years ago, thus reinforcing the notion that these social behavior tendencies are basic and profound in our natures. Women during pregnancy, and carrying small infants, could hardly do the running, hunting, and gathering as well as the male could. So, a mutually cooperative relation between male and female became essential. Monogamy, loyalty, and friendship all encourage our positive, helpful and, in a word, altruistic responses toward each other. As I said in my book, *The Neuroscience of Fair Play*, friendly, cooperative behaviors require brain mechanisms—hormonal, genetic, neural—that subsequently become available to support a wide variety of friendly, supportive behaviors that have nothing directly to do with sex or maternal behaviors. In evolutionary terms, that is, once the mechanisms required for male/female courtship and sexual behaviors are in place, they are at the service of more complex social relations of a positive, cooperative sort. And all of those friendly behaviors require social recognition.

Since identification of another as distinct from, or similar to oneself plays a major role in the chapter, I must ask: How exactly do we recognize others for who they are, as distinct from ourselves? We are beginning to piece together the molecular basis of social recognition through brain research on laboratory animals. As distinct from humans, whose visual and auditory capacities are so rich and strong, these laboratory animals tend to rely on smell. Because virtually all pheromones and other odors signal through basal forebrain pathways that lead to the amygdala, this collection of neurons once again comes into play. Pheromonal signals from the accessory olfactory bulb impact the medial nucleus of the amygdale, while volatile olfactory signals from the main olfactory bulb converge on the other amygdaloid cell groups.

Biologist Elena Choleris investigated these mechanisms when she was in my lab at Rockefeller University. Choleris, born in Italy and raised

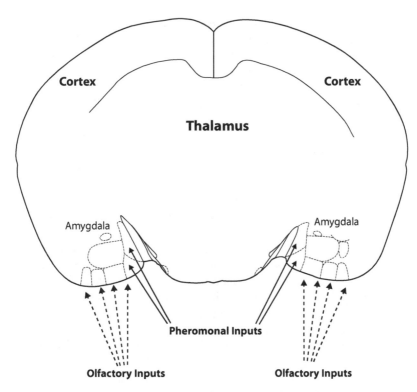

Pheromonal and olfactory inputs to the amygdala. Volatile odor signals from other animals are processed through the main olfactory bulb, whose neurons then project to and excite electrical activity in a variety of amygdaloid cell groups and adjacent cortex. Many pheromones are quite nonvolatile. Their signaling is processed through the accessory olfactory bulb, whose axons tend to project only to the most medial cell groups in the amygdala.

in Greece, now works in Canada, but has a Broadway-bound sense of humor. One day in my lab, she pointed at a drawing of the very small Y chromosome and wrote on it: "So small and yet so dangerous!" Elena studied female mice because among lower mammalian species, such as the rodent species, females form social bonds much more readily than males. These laboratory rodents, having poor vision and being nocturnal in their habits, depend largely on olfaction and on pheromones for their social recognition capacities. Choleris analyzed the abilities of female mice to recognize other females using precise assays that told us not only how the test females got used to females they recognized—and knew that those intruder females did not represent any threat—but also told us how the test females would reawaken their investigative responses when a new, strange female was introduced. We studied genetically normal mice, and then compared their social recognition performance to that of their littermates in which the oxytocin gene had been "knocked out," or either of two sex hormone receptor genes had been knocked out—estrogen receptor-alpha, or estrogen receptor-beta. We manipulated these sex hormone receptor genes because in lower mammals, many aspects of social behavior occur in hormonally-dependent settings connected with reproduction. Choleris found that "knocking out" any one of those three genes would significantly reduce all aspects of social recognition.

Choleris put those results together with a long and strong literature on the molecular actions of estrogens in the brain, and her understanding of how mice use olfaction and pheromones to recognize each other. She came up with a coherent story that links social behaviors to reproduction. The female mouse's ovaries secrete estrogens as the animal is getting ready to ovulate. Circulating in the blood, these estrogens are retained by neurons in the hypothalamus, and in those neurons the estrogens turn on the gene that codes for oxytocin (OT). The elevated levels of OT are transported to the amygdala. At the same time, the estrogens, having circulated in the blood, also are retained by neurons in the amygdala, where they turn on the gene for the oxytocin receptor (OTR). Thus, elevated levels of OTR are there, ready to receive and react to the OT transported from the hypothalamus. Choleris emphasized that *concurrent* expression of these various genes in their respective different *locations* in the forebrain would be crucial for social recognition to work correctly. The fact that these molecular events take place in the amyygdala is important for two reasons.

First, it is to the amygdala that olfactory and pheromonal signals signal, providing the basis for social recognition. Second, it is precisely in the amygdala where OT, working through OTR, fosters increased social recognition. Jennifer Ferguson, a graduate student working with Thomas Insel when he was at Emory University, found that microinjections of OT to the amygdala improved social recognition. Conversely, Choleris and I used a special molecular trick called *antisense DNA* to block gene

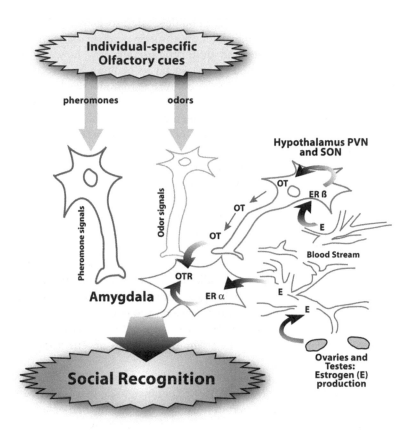

According to the results of Elena Choleris in our lab at Rockefeller University, estrogens (E) circulate in the blood and enter the brain. Arriving at the hypothalamic cell group, the paraventricular nucleus (PVN), they bind to estrogen receptor (ER)-beta and as a result stimulate transcription from the oxytocin (OT) gene. Some of those cells send axons to the amygdala. Meanwhile, estrogens have arrived at the amygdala and bind to ER-alpha, and thus can stimulate transcription from the OT receptor gene (OTR). When OT has bound to OTR, these amygdaloid neurons more efficiently process pheromonal and odor signals from other mice, producing better social recognition of other mice. Put these molecular mechanisms together, and you have a coherent theory of how estrogens foster friendly behaviors among laboratory animals that rely heavily on pheromonal and odiferous social signals. I propose that these mechanisms have been conserved even into the human brain, and are now combined with a variety of other regulatory mechanisms.

expression for OTR in the amygdale, and decreased social recognition. As a result, from molecular chemical details, through neuroanatomy, through animal behavior, we understand quite comprehensively how OT and OTR operating in the amygdala foster social recognition in mice. Additionally, we see that it operates in the context of reproductive hormones. I have argued in my MIT book, *Drive,* that these primitive molecular and neuroanatomical relations have been retained in the human brain, and operate in much the same way. But, of course, our social relationships depend on myriad cultural habits and customs overlaying the primitive, sexy drives, and depending on our newly evolved cerebral cortex, even as, neuroanatomically, the human cerebral cortex overlays the human amygdala.

I note that Elena Choleris' model of social recognition, even in these laboratory mice, does not invoke a simple-minded statement that claims "one gene/one behavior"—i.e., turn on gene A and behavior B pops out. Newspaper reporters sometimes talk that way, because classically in genetics, decades ago, George Beadle and Edward Tatum, studying the fungus Neurospora, won the Nobel prize for their "one gene/one enzyme" concept. But modern neuroscientists have moved beyond that. In my lab, to explain mechanisms for behaviors that are different between males and females, we have shown that *patterns* of genes govern *patterns* of behaviors.

Nor does Choleris try to wrap up the explanation of social recognition and altruism in a single gene. Her reluctance to do that is ratified by a mathematical approach reported in *Nature* in 2009 by mathematicians Vincent Jansen and Minus van Baalen, at the University of London. By simulating what would happen in social conflict, setting cooperative instincts against selfish instincts, and by assuming that social recognition and altruism are always inherited together, Jansen and van Baalen mathematically tested the implications of the simplest idea: that social recognition and altruism both depended exclusively on the same gene. Their calculations revealed that that coinheritance would lead to great instability. Social cooperation would bounce from absent to present, depending on whether or not that single gene existed in an individual, in a manner that would not be sufficient to support a normal society, animal or human. If, instead, they assumed that social recognition and altruism were caused by "loosely coupled separate genes," then the potential would increase for the development of a variety of recognizable features across the population that would greatly foster altruistic behavior. Jansen and van Baalen's conclusion is important because it provides a genetic mechanism by which people could recognize each other as altruistic and behave appropriately even if they are not in the same family or otherwise related to each other.

Biologists have long ago grasped how social behavior networks have evolved among all animals with backbones, vertebrates, encompassing a range of animals from fish through the types of laboratory mammals whose maternal behavior I described previously. Even in fish, neurobiologist James Goodson and his colleagues at the University of California at San Diego can discern hypothalamic/amygdala relations of the sort that Elena Choleris identified. Goodson also sees that in fish, social behaviors in the form of vocalizations are linked to their requirements for reproduction. Birdsong provides us with another obvious example. In many bird species, males sing much more than females and use their songs to control their territories and attract females. But for the main arguments of this book, extending the story to higher mammals such as nonhuman primates and thence to humans is most important. In the words of Robert Axelrod, computer scientist at the University of Michigan, once the genes for cooperative behavior have evolved, natural selection of optimal social behaviors will operate and will produce "strategies that base cooperative behavior on cues from the environment."

The amygdala, highlighted in the following figure, is a sexually differentiated part of the brain. I note that events in the female amygdala that dispose the animal toward friendly social behaviors can be distinguished sharply from the male amygdala that I'll describe later. That is, when things go wrong in the male amygdala you can expect autistic behavior to emerge.

Friendship

"I love him like a brother!" "She's been like a daughter to me all of these years." We all have heard these expressions. I think that loving, supportive relations typical of stable sex partners and families blend into our feelings for friends in general. As an important consequence of this idea, mechanisms of sex behaviors and maternal behaviors discussed earlier tell us a lot about mechanisms for the positive, friendly, ethical behaviors that conform to the golden rule.

(SOCIAL) TRAUMA

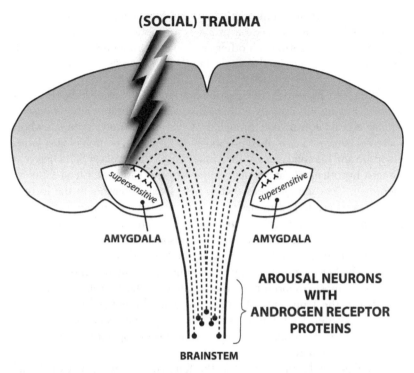

Why do males have autism so much more frequently than females? About 80% of diagnoses of autism are boys, and 90% of Asperger's diagnoses. This drawing encompasses a theory to explain the preponderance of males among autistics. Testosterone revs up activity amongst arousal-causing neurons in the lower brainstem, neurons whose axons are known to impact the amygdala. At the amygdala, judging from the findings of Benno Roozendaal and James McGaugh at the University of California Irvine, these arousal-related inputs are essential for the production of fear and anxiety. In boys with a supersensitive amygdala, traumatic stimuli before or after birth, especially stimuli connected with other people, cause these boys to avoid the source of the trauma, other people. According to my approach, little boys who will be autistic don't avoid social contact because they don't know how to be social, they avoid social contact because they know they would be made anxious by it.

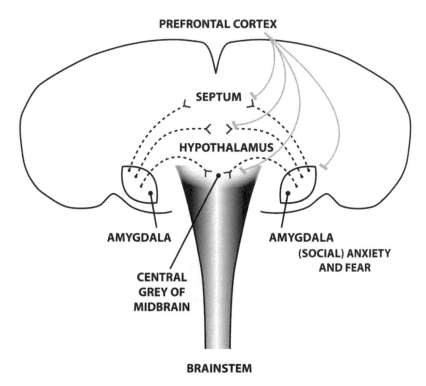

Axons from the prefrontal cortex dampen fear by reducing exitability in the amygdala, and counteracting effects of amygdala outputs to other brain regions.

In animal and human life, what has to happen for ordinary friendliness? First we must learn the other's identity, using the mechanisms previously described, so that we can remember and recognize that person. Second, assuming that the other person has not harmed us—no danger there—we recognize him or her as a friend. Then, we are permitted to blur that person's features, his identity with our own, and we develop empathy for that person. As I concluded in my *Neuroscience of Fair Play*, we can "love that person as ourselves."

How does friendship work, in terms of human brain mechanisms? Well, females and males are not the same. Some of the investigations of the human brain that neuroscientists carry out these days are neuroanatomical.

Tania Singer and her team, at the University College of London, are using fMRI to reveal some of the nerve cell groups whose activities are correlated with empathy of one human for another. They asked volunteers to play a game in which employees of the experimenters either played fairly or unfairly. Then, they measured brain activity in these same volunteers while they observed those employees receiving pain. Both male and female volunteers exhibited activation in parts of the brain that signal pain when they watched fair-playing employees receive pain. This was a neural sign of empathy, because the volunteers could sympathize with fair-playing employees. However, when unfair-playing employees received pain, this empathic neural response was significantly reduced in males, but not females. Thus, females showed brain responses associated with empathy regardless of their moral valuation of the employees' social behavior, whereas men's brain responses depended on how fairly the employee had played in Singer's experimental game.

Genes expressed in neurons in the forebrain help to support the kind of neuronal activity necessary for the feelings and behaviors studied by Tania Singer. Barry Keverne, head of the Department of Animal Behaviour at the University of Cambridge, has launched the discussion in terms of the evolutionary history leading to social bonding among members of monogamous species. Keverne states that many nonhuman primates would not survive if they depended on the bond between mother and infant alone. Although mechanistically, as I am arguing hscale social bonding owes its working parts to those genetic and neural steps that permit maternal behavior, they are not enough by themselves. In addition, there must be extended family relationships that permit living in larger social groups. Keverne points out that as primate brains grew during evolutionary history, and the distance senses like vision and hearing gained in range and power, the role of olfaction in social recognition and bonding declined in importance. Nonetheless, as anticipated from my discussion of the genes coding for OT and OTR in mice, Keverne emphasizes their importance in social recognition and bonding in humans as well. Looking to the future, the neuroanatomic and genetic explorations typified by Singer's and Keverne's work will have to be woven into comprehensive explanations of sex differences in friendly human behaviors.

The upshot of all of these brain mechanisms, including those we do not even know about yet, is to produce social behaviors with highly variable degrees of affability and tenderness. And these are not equal between women and men. Shelley Taylor, a professor of psychology at UCLA, and her colleague Laura Klein, now at Pennsylvania State University, summarized literature on the biology of behavior by characterizing women's social behaviors as "tend and befriend" behaviors, as opposed to males' behavior, which is more typified by the formation of dominance hierarchies. In Taylor's words, women are more likely then men, in response to stress, to protect offspring and turn to the social group for aid. From the point of view of hormone biology and nervous system biology, Taylor would theorize that friendly female behaviors depend heavily on estrogenic actions in the brain, especially as they enhance the behavioral effects of oxytocin. I agree with her summary, especially since it jibes well with Elena Choleris' work with animal brains. Androgenic hormones, higher in men, probably antagonize these estrogenic actions. In fact, Klein has reported that certain neurochemicals such as opioid peptides have opposite effects on the social behaviors of men and women. This field of study needs a lot more work, but already we can see elements of the neurochemical basis for sex differences in social behaviors.

All of these sex differences are heightened by stress. Under tense circumstances, men—especially young men—are more likely to think about fighting, while women are more likely to use conciliatory or negotiating tactics. Of course, all of these tendencies have a statistical character. Some women may be exceptionally combative, while some men may be exceptionally conciliatory. Nevertheless, the sex differences in social behaviors characterized by Shelley Taylor and her colleagues have drawn the attention of evolutionary biologists and social anthropologists.

Some evolutionary biologists would hypothesize that differences in affability and tenderness between women and men derive from the greater degree to which females invest their energies in parenting. As briefly touched upon earlier, the temperaments required for caring for tiny babies would carry over into attitudes displayed toward other adults, especially under stress. Other biologists would point toward the evolutionary advantages of social groups whose members display reciprocal altruism: "I will treat you just as I would want to be treated myself." In my *Neuroscience of Fair Play*, a book whose thinking has influenced my treatment of sex differences, I point out that the human statement of reciprocal altruism, the "Golden Rule" appears to be universal among the world's religions and can be explained in a parsimonious theory of cerebral cortical physiology. Scientists value "parsimony" in their theories, meaning that they do not want to make any self-serving assumptions about the systems they are trying to elucidate. In our case, therefore, I am pleased to claim that no unusual or special abilities of the central nervous system need be proposed in order to explain reciprocally altruistic behaviors.

In modern societies, women are taking positions of leadership that might not have been expected of them years ago. Women head nations and major corporations, and hold highly responsible positions in the military. Does that mean that we should predict, about them as individuals or about entire classes of women in general, that females will stop "tending and befriending," in Professor Taylor's words, and will become dominance oriented? I don't know. As stated earlier, the sex differences in social behaviors are statistical tendencies, not physical absolutes. If I had to guess, I would predict that some details of the styles of leadership demonstrated by women in power will differ substantially from those styles shown by some powerful men in the past. On the average, women will be less confrontational and will more efficiently seek consensus.

Christine Drea, at the Department of Biological Anthropology at Duke University, asked analogous questions about a nonhuman primate famous for female social dominance, the ringtailed lemur. Indeed, in terms of genital anatomy and neuroendocrine parameters, many female lemurs were "masculinized" and had an unusually high level of a steroid called *androstenedione*, a precursor of androgenic hormones. Females retain a privilege called "feeding priority" in which males defer to females, and they may display "overt aggression against males." This pattern of correlations between hormonal factors and behavioral status and aggression does, in Drea's words, "suggest a possible role for androgens in feminine development" in this species that shows female social dominance. However, I add two caveats that will head off any oversimplified inferences. First, we are talking about population phenomena here, statistical features that do not adequately support the conclusion that androgenic hormones lead to unusual features of neural and anatomical development, that in turn lead to dominant behavior in any individual female lemur. Secondly, among humans, the liberation of relatively simple behaviors that, in lower species, are slavishly dependent on hormones, has led to social behavior repertoires that (a) are less dependent on hormones and more dependent on cultural influences, and (b) are so finely graduated and articulated that considerations of one-dimensional "dominance" or "aggressiveness" are strictly yesterday's news. Current thinking tells us to pay attention to the details of social situations in which sex differences in power relations are playing out, and to respect large differences among individuals of each sex as they respond to those social situations.

As opposed to the tending and befriending that Shelley Taylor and others celebrate among women, the situation for men, historically, has been quite different. Not only do we have the competitive, hierarchically obsessed man, but we also have some little boys who actively avoid social interaction.

When Friendship Is Shunned—Autistic Males

If in some respects males are "opposite" to females in their social behaviors, then what, exactly, would we consider to be the opposite of the "tending and befriending" that girls often do so very well? Of course, one instinct that is opposite is the instinct toward aggression. But another type of opposite to females' friendliness is the complete *avoidance* of social interaction. In its extreme form, this avoidance of social interaction would come under the heading of autism. The many forms of this disease have led to the phrase Autism Spectrum Disorders (ASD). Even though autistic patients can seem totally out of touch, especially at events like parties, many are not intellectually impaired. One group of them, Asperger's syndrome patients, are so smart and are so good at concentrating on their favorite areas of study that Asperger called them "little professors." So the condition does not involve intelligence. It must be something else.

First attempts at modern neurobiological research about ASD have focused on the sensory systems we know best: vision and audition. Neurobiologists have asked, "Well, if the ASD patient does not seem to recognize a friendly face, or gaze at other peoples' faces, or respond to friendly words, then is that person's visual system okay? His oculomotor system? His auditory system?" However, at least so far, those sensory systems studies have not helped us at all. To date, no studies have indicated that deficits in autistic patients' sensory systems interfere with their ability to respond to faces and words. Instead, because prevalence of ASD is significantly higher among boys than girls, it is likely that important answers will lie not in sensory systems but in emotional systems. Another reason for thinking about emotional factors has to do with the sex difference.

More than 80% of ASD diagnoses are boys, and more than 90% of highly intelligent Asperger's diagnoses are boys. Not believing that this sex difference has to do with intellect or with sensory systems, I think it highly likely that ASD conditions have something to do with hormone effects on emotions. In order to explain how boys could be afflicted more than girls, I have to bring in the steroid hormone testosterone, acting through its special receptor, the androgen receptor (AR). According to my neuroscientific approach, testosterone acting in the very young boy's brain fosters higher levels of fear or anxiety in boys, subsequently to cause those boys to avoid social interactions in a manner typical of ASD patients. Their avoidance response has been learned so well that they simply fail to engage in social interactions. Here is how things work.

Throughout the boy's life, testosterone working through AR impacts primitive systems in the lower regions of the brain—primitive systems that arouse the entire central nervous system, notably a forebrain region called the *amygdala*. According to the results of behavioral neuroscientists James McGaugh at the University of California at Irvine, and Benno Roozendaal at Groningen University in the Netherlands, these brain arousal inputs to the amygdala are very important in priming amygdala cells involved in fear and anxiety. These inputs make the little boy supersensitive to a wide variety of prenatal and neonatal events that could cause anxiety. While scientists often use the word *trauma* in reference to a testosterone-primed, supersensitive boy, triggering events could be as mild as having a stern father or an absent mother (for instance, when she goes to the hospital to have her next child). Why some boys and not others? First, there are variations among individuals in prenatal and neonatal testosterone levels and, more subtly, in brain tissue sensitivity to testosterone. I'll discuss this later. Second, mothers' experiences during pregnancy vary widely, as do the comfort levels and skills of the newborn's parents. Crucial is the fact that McGaugh and Roozendaal, and a host of other scientists, have shown how arousing transmitters like norepinephrine or dopamine heighten amygdala neurons' ability to cause fear and anxiety.

Exactly how do amygdala neurons cause fear or anxiety? It is through their outputs to at least two places in the forebrain and one place in the midbrain that regulate those emotional events. In the forebrain, a virtual sliver of cells, a group tall and narrow just on the midline of the brain, is called the *septum*, and a tiny region of cells at the bottom of the brain just above the pituitary gland is called the *hypothalamus*. These both regulate such emotions. And in the midbrain, a small bunch of cells layered around the center of the brain, the midbrain central grey, is intimately connected with fear. If all of these zones receiving amygdala outputs are going full blast, fear and anxiety of all sorts, including some aspects of social anxiety, are to be expected.

The effects of these amygdala outputs can be dampened. From the very frontmost parts of our brains, in the prefrontal cortex, neurons projecting to these amygdaloid target zones and, indeed, to the amygdala itself, can reduce the fear-provoking effects of amygdala activity. Since the prefrontal cortex is responsible for the abilities that make us human, sophisticated thinking and decision making, I put forward the notion that cognitive retraining of boys with autistic symptoms, as would make use of their prefrontal cortical capacities, may work to reduce autistic symptoms. If this is true, noninvasive therapy, without drugs, could be envisioned to do the trick. Such cognitive retraining might well use the techniques of behavior modification: steadily increasing social exposure and connecting those social forays with rewards rather than with punishment should work. There is no reason that these testosterone-laden autistic boys, often smart and supersensitive, can not learn the pleasures of social engagement just as readily as they earlier were turned off social engagement.

And so, recognizing that all of us blend "typical female" and "typical male" social characteristics, or mixtures thereof, into our individual temperaments, I nevertheless believe the psychological evidence that in many situations females are more likely to be friendly and conciliatory, while males in similar situations are likely, on the average, to be less interactive.

The Story So Far

After all of this discussion about social capacities, I'll turn my attention in the next chapter to physical and psychological maladies that are suffered by the individual, male or female.

EIGHT

Pain and Suffering

Putting together all the mechanisms we neurobiologists have discovered for sex differences in brain and behavior, I've discussed several primitive behaviors that are so strongly sexually differentiated—mating, parenting, aggression, and certain social predispositions—that you would be convinced of the scientific power of my arguments. For these primitive behaviors, the conservation of neuronal systems during evolution tells us that many of the human brain mechanisms have been "left over" from animal brain mechanisms.

That's fine for normal behaviors. For normal behaviors, our central nervous systems are functioning well. But what about times when things are going miserably? Are you in pain? Under stress? If so, it will make a difference whether you are a typical male or a typical female.

Further, if you are female, you are much more likely to suffer from anorexia nervosa, a severe eating disorder that can lead to death. Depressed? In the United States, the ratio of female sufferers to males is about 2.5 to 1, and in Denmark, 6 to 1! And, if you have one of the serious fatigue disorders like chronic fatigue immune dysfunction syndrome (CFIDS), you're female by a ratio of about 8 to 1.

What about problems more often diagnosed in males? We have already covered extreme aggression and autism spectrum disorders, both more frequent in males. In this chapter I'll focus on another male-typical condition, attention deficit hyperactivity disorder (ADHD).

Quoting Thomas Insel, now head of the National Institute of Mental Health in Bethesda, Maryland: "It's pretty difficult to find any single factor that's more predictive for some of these diseases than gender."

Pain

Everyone, male or female, experiences pain. But special systems in the brain for reducing pain, and for preventing pain from interfering with other important behavioral responses, are different in males and females. There are, in short, sex differences both in the reception of pain and the suppression of pain.

To quote pain expert Karen Berkeley of Florida State University, "the burden of pain is greater, more varied and more variable for women than for men." She worked on a consensus among pain researchers who considered more than 80 different kinds of pain, and these researchers agreed that female prevalence in painful disorders was more than twice male prevalence. Yes, of course both men and women had acute tension headaches and toothaches. But many common disorders, such as *tic douloureux* (painful swelling of the nerve that delivers feeling to the face), migraine headaches with auras, chronic tension headaches, and pains of psychological origin were more often experienced by women.

How do these sex differences in painful experiences arise? Some of the mechanisms are surprising. For example, it turns out that the lower one's blood pressure is, the greater his or her sensitivity to pain will be. Therefore, in some studies, women's greater sensitivity to certain kinds of painful stimuli could be accounted for by women's lower than average blood pressure. Another difference derives from women having uteri, disorders of which can radiate pain widely to muscles.

Other mechanisms you would suspect. Women have two X chromosomes, and certain genes on the X chromosomes, in their recessive alleles on both X's, lead to a fair number of painful diseases that men simply would not get.

And then there are women's hormones. The literature on this topic is confusing, but a few clear stories stand out. Alan Gintzler, at the medical school of the State University of New York, asked how a woman can endure the excruciating experience of giving birth. It turns out that in her spinal cord she has her own, built-in pain control mechanism that is sensitive to the hormones of pregnancy and helps her do the job. This system demands to be run in just the right manner. Both estrogens and progesterone must be on board; neither alone will do. Under the influence of these two sex hormones, two different kinds of opium-like peptides come into play: those that stimulate so-called kappa opioid receptors on nerve cells in the spinal cord, and those that stimulate delta opioid receptors on nerve cells in the spinal cord. Neither receptor type alone will do. As a

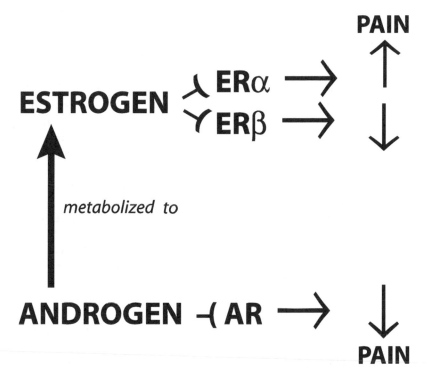

The effects of estrogens on pain depend on which estrogen receptor (ER) gene product they bind to: bind to ER-alpha, increased pain; ER-beta, decreased. If androgens are not metabolized to estrogens, and bind to the androgen receptor (AR), the predominant effect is a decrease in pain.

result of the actions of these opium-like peptides, a woman's discomfort during labor is reduced.

Outside of pregnancy, however, a woman's experience of pain becomes more challenging. Her estrogenic hormones have different effects on pain depending on which estrogen receptor they bind to. If they bind to estrogen receptor-alpha (ERα), they increase pain. If instead they bind to ER-beta, they will decrease pain. They can act at many places in the nervous system to have these effects, starting with the sensory receptor neurons themselves, in the spinal cord and in the brainstem. Anne Murphy, neuroanatomist at Georgia State University, has emphasized the importance of estrogen receptors in one particular part of the brainstem, the midbrain central grey. In all the vertebrate species I know about, pain pathways ascending in the nervous system do encounter the midbrain central grey.

These hormone receptors in the midbrain central grey lead us directly to another source of sex differences in pain: the dampening of the painful experience by systems in the brain. In our hindbrain, just above the spinal cord and in the midbrain central grey, neurons treated in just the right way can significantly reduce pain. They cause partial analgesia. Anne Murphy and her colleagues have built upon the work of behavioral neuroscientist Richard Bodnar, at Queens College, to demonstrate the neurochemical basis of this analgesia. As you know, morphine can reduce pain. But males feel this reduction of pain, this analgesia, much more than females. Bodnar's group demonstrated that the sex difference in morphine-caused analgesia is due to sex differences in the brain and, in fact, in the midbrain central grey. Opium-like peptides operating through mu-type opioid receptors in this part of the midbrain were more effective at reducing pain in males than in females. This was due to the actions of testicular hormones on the brain early in brain development. In two landmark studies, Murphy and her team found that morphine activated more of the most important central grey neurons in males than in females—these were neurons that sent their signals to an important pain-controlling region—and that this functional sex difference is due to greater mu-opioid receptor expression in males.

When all is said and done, this sex difference in pain suppression is even larger than the sex difference in the initial thresholds for pain. One lesson I take away from this scientific work says that the very best medicines for controlling pain are likely to be different for females compared to males.

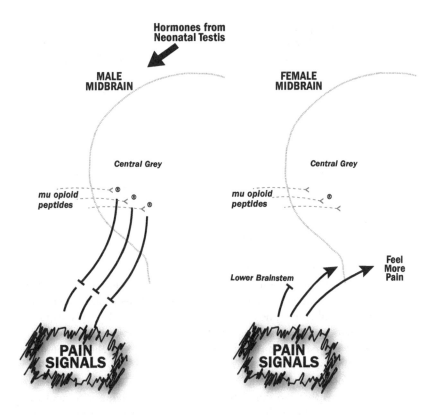

Males can suppress pain by the actions of mu-opioid peptides better than can females. Pain signals ascending through pathways from the spinal cord or the lower brainstem have a big impact on a portion of the midbrain called the *central grey* (because it is right in the middle of the midbrain). Deriving from the chemistry of opium, neuropeptides that target the mu-opioid receptors resident in the central grey are significantly more abundant in the male brain than in the female. Their actions dampen pain signaling. As a result, females, on average, will feel more pain through this route.

Professor Berkeley emphasizes that the experience of pain amounts to a perceptual and an emotional phenomenon, not just a simple physiological response to a sensory stimulus. Knowing that allows me to appreciate that cultural and social influences—influences that, for example, encourage men to be stoic, to "tough it out," to "suck it up"—may play some role with respect to sex differences in pain. To paraphrase Berkeley, some of these cultural influences may "lead women to appreciate and do something about threat at an earlier stage than men; i.e., women are generally smarter about danger than men."

Female Suffering

Anorexia

Anorexia literally means "without appetite." Women are diagnosed with anorexia nervosa almost ten times as often as men. Psychologically, anorexics fear becoming fat. Then, semistarved, they are vulnerable to irritability, social withdrawal, and depression. They become dehydrated, their skin becomes dry, their nails brittle, their blood anemic. They lose their periods and they are subject to heart abnormalities.

While previously, anorexia was thought of as a purely psychiatric disorder involving an obsessive-compulsive concentration on body image, biological factors are now understood to play a role as well. Exacerbating the psychological causes, hormonal changes can powerfully predispose a young woman to anorexia. It is usually seen in adolescent girls, when estrogen levels are rising in the blood. This leads me to believe that at the very beginning of anorexia nervosa, young women with hypersensitivity in one part of the hypothalamus, the ventromedial nucleus, are led by estrogen actions there to think they have "eaten enough." The neuroscientific literature tells us that estrogens, acting on this part of the hypothalamus, cause a reduction in food intake. An alternative explanation, not exclusive of the neurobiological factor, is that much of anorexia is culturally determined—girls looking at supermodels and worrying about their own body images—and that anorexics can be very hungry but fight off the urge to eat.

The ventromedial nucleus of the hypothalamus provides the regulation of reproduction in the light of nutritional needs, and the regulation of nutrition in the light of reproductive needs. A female laboratory animal will not try to mate if she is vastly undernourished. An animal that becomes pregnant, and will soon have a litter to nourish, will increase her food and water intake. Estrogens cooperate with hormones from the gut to achieve this balance between reproduction and nutrition in a biologically and medically adaptive manner.

But when estrogenic sensitivity is abnormal in the ventromedial hypothalamus of some individuals, troubles arise with respect to eating, and those troubles can go in both directions. Consider an "estrogen balance" hypothesis. In some girls, a sudden increase in estrogenic tone at the beginning of puberty will, I know from my own experiments, suddenly turn up activity in neurons of the ventromedial hypothalamus. Increased activity of these neurons has been proven many times to decrease food intake, and these girls will, as a consequence, eat less—perhaps to the point of becoming ill. Troubles wait for us in the other direction as well. Complete insensitivity to estrogens in ventromedial hypothalamic neurons will reduce the activity of these neurons and cause such individuals to eat more. Sergei Musatov, virologist at Cornell Medical School and cooperating with my lab, used a viral vector carrying a "killer" molecular sequence that would specifically destroy estrogen receptors in ventromedial hypothalamic neurons of female mice. In the complete absence of hypothalamic estrogen sensitivity in these animals, we saw symptoms of the dangerous "metabolic syndrome." The animals thus affected would eat more, have no increase of body heat after eating, exercise less, have abnormally sluggish responses to blood sugar, and as a result of all these changes, suffer increases in body weight and body fat. In other words, too great a sensitivity to pubertal increases in estrogen—anorexia. Too small—metabolic syndrome. In essence, women need a finely tuned balance of sensitivity to estrogen levels and their changes.

Stress and Anxiety

Elizabeth Young, a medical doctor at the University of Michigan who has long studied stress disorders, discusses both the easy aspects and the hardest aspects of studying stress. On the one hand, stress clearly causes the brain to release a tiny fragment of protein called *corticotropin releasing factor* (CRF) into the pituitary gland, telling the pituitary gland to release corticotropin into the blood, thus to stimulate the release of stress hormones like cortisol from the adrenal gland. In parallel with the release of stress hormones, our autonomic nervous systems cause our hearts to pound and our skin to sweat. Those are the easy aspects. The hardest aspect of studying stress? Environmental conditions that cause stress differ greatly from each other, may contain several separate factors and, obviously, are most complex for human beings.

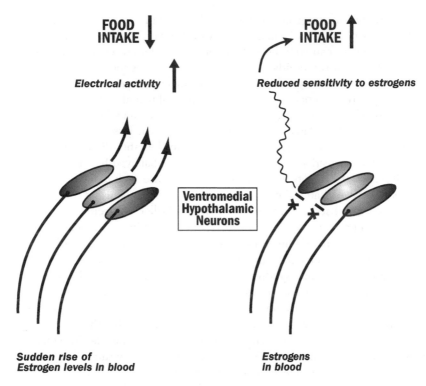

FOOD INTAKE ↓

Electrical activity ↑

FOOD INTAKE ↑

Reduced sensitivity to estrogens

Ventromedial Hypothalamic Neurons

Sudden rise of Estrogen levels in blood

Estrogens in blood

The ventromedial nucleus of the hypothalamus contains neurons that are extremely important for the regulation of food intake. In particular, they help to manage the relationship between nutrition and reproductive functions, especially important for women before and during their care of babies. Sudden increases of estrogen levels will increase the electrical activity of these hypothalamic neurons, and should decrease food intake. An abnormal sensitivity to this sudden rise of estrogens at the beginning of puberty could contribute to anorexia nervosa. Conversely, Sergei Musatov, a virologist working at our lab and Cornell Medical School, helped us use a viral vector to reduce the expression of the estrogen receptor gene in the ventromedial nucleus of the hypothalamus, with the result of increased food intake and body weight.

In laboratory rats, the release of stress hormones like cortisol from the adrenal gland following stress is greater in females than in males. In fact, females' adrenal glands, organs that produce the stress hormones, are larger than those of males. Chronic mild stress interferes more with

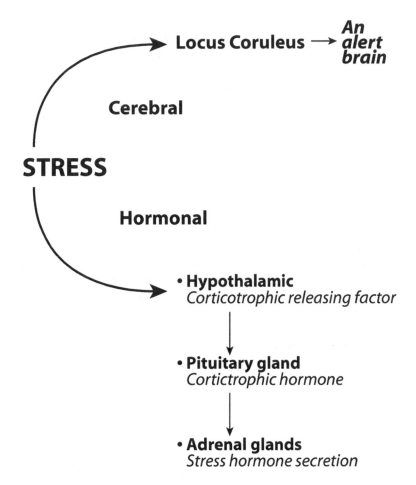

Stress triggers both hormonal (bottom half of drawing) and neuronal (an "alert brain") responses. Regarding the neuronal responses, for example, neurons in the locus coeruleus use their axons to ship the arousing neurotransmitter norepinephrine to large regions of the forebrain.

females' normal activities and with their appreciation of positive stimuli, such as when they are offered sucrose-loaded water. Across the board, says the expert Jaak Panksepp at Washington State University in Pullman, data from laboratory animals "overwhelmingly indicate that females show more intense fear responses than males." This holds true for women as well. Women react more strongly to stimuli that are emotionally negative, and even to the anticipation of negative stimuli.

At least three different sets of mechanisms cause women's greater responses to stress. First, their increased release of stress hormones from the adrenal gland. Under the influence of estrogens, both the magnitude and the duration of the female's cortisol responses to stress are greater. In addition, the responses of brainstem neurons that tell the autonomic nervous system to react with changes in our hearts, stomachs, breathing, and sweat glands are greater in females. Third is the mechanism of CRF acting directly in the brain. Recent research by Tracy Bale and her laboratory members at the University of Pennsylvania has directed our attention to a particular class of neuropeptide receptors in the brain. While CRF is the neuropeptide that tells the pituitary to tell the adrenal to pour out stress hormones, CRF also acts on other neurons in the brain. When Tracy Bale genetically engineered mice so that they would lack CRF receptors, recovery from stress was delayed and the mice were more anxious. Furthermore, the expression of the gene that produces CRF is increased by estrogens, female sex hormones.

These sex differences in stress responses have implications for several diseases, including anxiety. Many more women suffer from anxiety disorders than men, according to the authority Margaret Altemus at Cornell University Medical School. By her definition, for example, "Generalized anxiety disorder is uncontrollable worrying about multiple problems." The problems are real, but the amount of worrying is out of proportion to their importance or probability. Altemus lists additional symptoms: "muscle tension, fatigue, insomnia, restlessness, poor concentration and irritability." Presumably, some of the sex differences I mentioned when discussing responses to stress contribute to the prevalence of anxiety disorders among women, and help to explain womens' greater psychological fear responses.

Women who suffer anxiety and chronic stress are particularly susceptible to getting depressed. The more frequent onset of depression in women has drawn the attention of large numbers of medical researchers. Luckily, the five major aspects of depression can be measured in laboratory animals, namely: (1) The relative inability to initiate behavior (behavioral inhibition); (2) the loss of pleasure in formerly pleasurable things and activities (anhedonia); (3) the disruption of normal daily (circadian) rhythms;

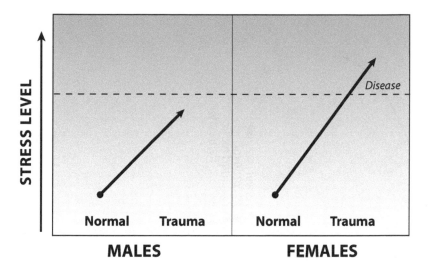

Thinking by Tracy Bale, molecular neuroendocrinologist at the University of Pennsylvania, and other stress physiologists, suggests that some traumas that increase stress levels in males, will do so to a greater extent in females, even to the point of causing symptoms of disease in these females.

(4) anxiety; and (5) sleep disturbances. All of these symptoms tell of problems with fundamental arousal pathways in the brain. As a result of these studies with laboratory animals, and subsequent clinical trials with depressed patients, some of the best current medicines that have been developed for treating depression are chemicals that heighten activity in arousal pathways of the brain. For example, they heighten the amount of arousal-regulating transmitters norepinephrine, dopamine, and serotonin in synaptic clefts by preventing them from being soaked up by the very neurons that had released them. Along with nonpharmacologic approaches—exercise, talk therapy, etc.—these drugs that heighten fundamental brain arousal should help a person beat her depression.

Fatigue Syndromes

Sometimes, clinical investigations permit mysterious and vague illnesses to emerge from the fog of medical misunderstanding and into an era when they can be dealt with directly and scientifically. Two such types of

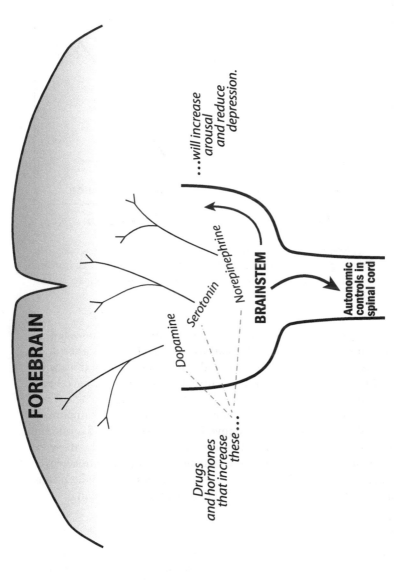

FOREBRAIN

Dopamine

Serotonin

Norepinephrine

Drugs and hormones that increase these...

...will increase arousal and reduce depression.

BRAINSTEM

Autonomic controls in spinal cord

Examples of neurochemical systems in the brainstem that use specific neurochemicals to regulate arousal of the brain. As a result, they influence the ability of a patient to initiate active behaviors. While projections from these brainstem systems to the forebrain, thus to heighten a patient's mood, are usually thought of as carrying the main effects of drugs that elevate dopamine, serotonin, and/or norepinephrine, projections from these brainstem systems to the spinal cord are probably very important, as well.

syndromes disable women much more frequently than they do men. One of them, Chronic Fatigue Syndrome (CFS), was long dismissed as a product of feminine hysteria. Now, the Center for Disease Control in Atlanta recognizes CFS as a valid medical syndrome, even though we cannot yet cure it in most cases. I can most easily describe it by presenting an individual case.

One patient I know, R.N., contracted the symptoms in her late 20s, a typical time for CFS symptoms to strike a young woman. At certain times she would just "run out of steam," feeling exhausted and run down. She would "hit the wall." At such times she not only would have no physical or mental energy, but she also would suffer muscle pains, weakness, and a feeling of hopelessness. She spoke in a weak voice and her breathing was labored. She would feel beaten down, overwhelmed, depressed. Her sleep was badly disturbed, to the point that she would have to nap during the day to survive. The last time I spoke with her she was too tired to talk, too tired to give voice to her chronic condition. Her experience is typical of Chronic Fatigue sufferers.

The other syndrome, Fibromyalgia Syndrome (FMS) is characterized by widespread muscular and skeletal pain, morning stiffness, headaches, sleep disturbances, and stomach problems. FMS sufferers, studied by neuroscientist Jon Zubieta at the University of Michigan, have reduced amounts of opioid peptide receptors in several regions of the brain that process pain signaling, thus offering a potential explanation for some of these patients' discomforts. I talked with a sufferer, a middle-aged teacher I'll name O.R., who had the type of chronic pain and consequent fatigue that led her to become depressed and to lose interest in the things she used to enjoy. O.R.'s self-esteem was down around her shoelaces, and she felt she had "lost part of her soul." O.R. said that "nothing was any fun." In her words, the pain "has taken over my life." Her social life was ruined. Although her sleep was okay, the sleep of others in her support group was disturbed. And, the fact that stress exacerbated her symptoms did not help.

Over the years, CFS and FMS began to be viewed as autoimmune syndromes, leading to CFS being renamed CFIDS (Chronic Fatigue Immune Dysfunction Syndrome). An autoimmune syndrome is defined as a condition in which a patient's immune defenses turn back against her and attack some of her own proteins. Exactly what is being attacked in CFIDS or FMS has not yet been determined. Moreover, autoimmune

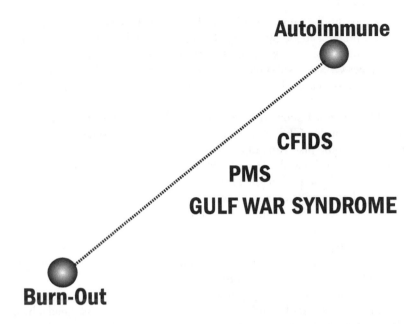

Autoimmune

CFIDS
PMS
GULF WAR SYNDROME

Burn-Out

There are syndromes that are purely autoimmune diseases, and there are syndromes that result purely from "burn-out." However, one speculation states that certain fatigue states— e.g., chronic fatigue syndrome, fibromyalgia syndrome and Gulf War syndrome—lie on a line that joins autoimmune and burn-out conditions, and that these fatigue states may be caused by a combination of autoimmune and burn-out determinants.

symptoms may not explain everything about these fatigue syndromes; for example, some doctors point to a role for chronic stress and subsequent mental and physical "burnout."

Across the board, females suffer autoimmune disorders more frequently than males. This subject needs a lot of research.

How do these symptoms come about, and why are females more susceptible than males? Unfortunately, there are all too many answers. Different mechanisms may work in combination with each other, and these combinations may form differently in different individuals. One huge set of mechanisms resides in the fact that estrogens and androgens have their receptors in immune cells, and affect immune cells differently. Other causes have to do with different reactions of women and men to different kinds of stress, as discussed previously. And then, at least three kinds of immune cells are resident in the brain: microglia, dendritic cells, and mast cells. These are pharmacological "bombshells." Mast cells, for example, can secrete histamine, prostaglandins, and inflammatory compounds, all of which could wreak havoc on nearby neurons. How do they respond to a woman's state of mind, and what effects do their reactions have? Finally, the effects of a woman's cultural surroundings—the pressures placed upon her by family expectations, her birth order, her educational status and countless other environmental details—must not be disregarded during our "high tech" discussion of endocrinology and the

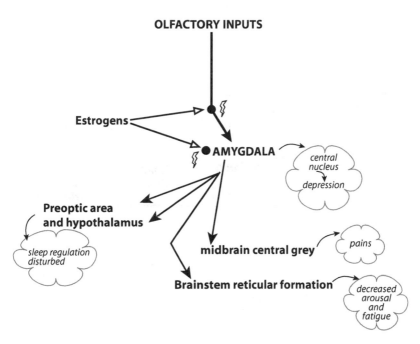

Multiple chemical sensitivity (MCS) entails extremely unpleasant responses to odors including pain, dizziness and feelings of nausea. MCS is highly correlated with chronic fatigue syndrome, which can include depression and sleep disturbances. Both of these syndromes are much more common among young women than among young men. I theorize that during adolescence in some females, estrogenic hormone actions on amygdala neurons—neurons that receive huge amounts of olfactory inputs—cause those neurons to be supersensitive (lightning bolts) and thus to cause the overreactions.

nervous system. Responses of friends and family to a young woman's early expression of distress may help determine the length and severity of her symptoms of fatigue.

Both CFIDS patients and FMS patients frequently react very strongly to odors, be they strong perfumes and aftershave lotions, or sweet-smelling products and scented detergents. Sufferers may become nauseous, dizzy, get headaches and feel faint. This pattern of multiple chemical sensitivities (MCS), highly correlated with CFIDS and FMS, points our thinking directly at parts of the forebrain that receive odors. The high sensitivity to odors, really oversensitivity, stimulates my neuroscientific theory of what MCS is all about, a theory illustrated here in outline form.

Male Suffering

Gulf War Syndrome

Another type of fatigue syndrome, Gulf War Syndrome, has symptoms very similar to CFIDS and FMS but is much more frequent in men than in women. I am amazed about the similarity of symptoms among these

	CHRONIC FATIGUE SYNDROME	FIBROMYALGIA SYNDROME	GULF WAR SYNDROME
Fatigue	✔	✔	✔
Muscle/joint pains	✔	✔	✔
Sleep problems	✔	✔	✔
Difficulty concentrating	✔	✔	✔
Headache	✔	✔	✔
Depression	✔	✔	✔

Many symptoms are common among Chronic Fatigue Syndrome and Fibromyalgia Syndrome sufferers, primarily women, as well as among Gulf War Syndrome sufferers, primarily men.

three syndromes, because the backgrounds and the experiences of the sufferers are so different. At first, like CFIDS and FMS, the Gulf War Syndrome lacked "face value." Now we realize the result of the pathology is the same, but the triggers are different and may depend on social experiences that are different between the sexes as much as on the underlying biology.

I have already talked about abnormal aggression and about autism, both more frequently suffered by males. Still another condition that is prevalent in males is attention deficit hyperactivity disorder (ADHD). Boys are diagnosed with ADHD about three times as frequently as girls.

Attention Deficit Hyperactivity Disorder

Children with ADHD exhibit an inability to stay still, and have a hard time listening, because they are impulsive and easily distracted. The balance among these symptoms—hyperactive, inattentive and impulsive—varies from kid to kid.

The causes of ADHD are unknown. While it could result from a variety of initial causes, we certainly should seek factors that explain the sex difference. Therefore, sex hormones that work through hormone receptors that boys have much more strongly than girls might be the way to go. Androgen receptors fill the bill. Their presence in skeletal muscle and in the lower brainstem reticular formation would, theoretically, explain how growing boys might be more susceptible to ADHD.

ADHD can be treated in two ways: medical and behavioral. The most typical medical approach, one that uses methylphenidate (Ritalin) baffles me, because its effects in the brain should be exactly the opposite of what we would desire. When Ritalin works, its effectiveness has been related to its ability to increase levels of the neurotransmitter dopamine (DA) in synaptic clefts in the forebrain. But DA release causes directed motor acts towards salient stimuli. So, how DA elevation can help ADHD, I don't know, and find the basic neuroscience of ADHD to be stuck in a very unsatisfactory state.

Behavioral therapy of ADHD includes talk therapy and behavioral modification techniques without any medication. These techniques have also included parent training, consultations in school, and special summer treatment programs. Finally, I was surprised to read that combinations

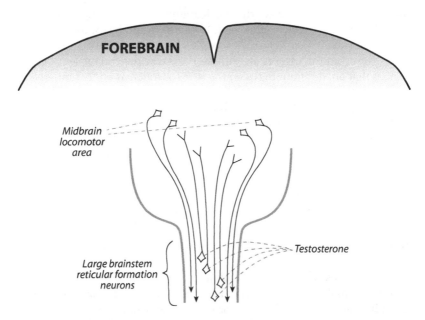

Why do boys seem to get ADHD more frequently than girls? One possible answer is that testosterone, having bound to androgen receptor proteins in large, powerful reticular formation neurons in the brainstems of susceptible boys, activate these neurons. Since these brainstem reticular formation neurons are known to project to the midbrain locomotor area, they can render those locomotor-facilitating neurons very excitable, thus causing considerable restlessness.

of treatments, behavioral and medical, did not necessarily yield better outcomes than either one alone. Put simply, ADHD research has produced a large volume of bewildering results that sometimes contradict each other. I am beginning to wonder whether many of the boys so diagnosed are absolutely normal—muscular and vigorous—and that our schools are simply ill-equipped to deal with them. Forcing healthy, active boys to sit still for too long probably denies the value of an evolutionary process that led to considerable large muscle strength in males, used in farming and hunting. So let these boys run!

The Story So Far

We have just discussed how males and females differ in their reception and suppression of pain. Genes and hormones both play a part. So does the brain, and the way men and women react to stress. Now that we have covered the effects of hormones on sexual and nonsexual behavior, we can move on to the topic of neonatal and pubertal environment's and their power to influence sexually differentiated behaviors.

NINE

Perilous Times—Newborns and Adolescents

Up until now I've been emphasizing biological events—genetic, neural and hormonal—as primary for driving behaviors that are different between men and women. Finally, I get to talk at greater length about two times of life during which environmental realities press inward on the individual, and influence the development of precisely those emotional behaviors that are sexually differentiated. These are the times close to a baby's birth, and the years during which boys and girls go through puberty.

One reason neonatal and pubertal environments have the power to influence sexually differentiated behaviors centers on stress. Stressful environments during these periods of changing hormone levels can crush the normal development of some behaviors and encourage others. Makes sense—in lab animals, the neonatal period is precisely when sex hormones cause sexual differentiation of the brain, and in the pubertal period the sex differences in depression and aggression, coupled with hormonal fluctuations, become visible. In these two periods of vulnerability, as hormones and hormone-dependent behaviors change rapidly, stress can impact our nervous systems with special force.

Birth

Failures of parental care will lead to stress for the baby, and stress for the baby will have consequences for his emotional disposition the rest of his life.

The late Harry Harlow, a professor of psychology at the University of Wisconsin, did experiments on the brains and behaviors of rhesus monkeys as a way of exploring the stressful effects of poor parental care on the social and emotional development of the offspring. He broke open this field of work by dividing baby monkeys into two groups. One group had normal maternal care. The other group had excellent care—they were well fed, and were not harmed in any way—but they were raised with mechanical "mother substitutes." As they grew up, instead of enjoying vigorous play in social groups of monkeys, the babies were withdrawn and depressed. In his classic book, *Learning to Love,* Harlow says, "The first of the affectional systems is maternal love, the love of the mother for her child. The second is infant love, the love of the infant for the mother...." If this connection fails to form, the child will have a significantly poorer chance of a rich and healthy social and emotional life. Scientific generations of workers have extended Harlow's work in experimental and clinical settings, and have continued to chart these consequences of difficulties during the neonatal period.

While most scholars have focused on mother/infant relations, the father is important, too. Social psychologists and psychotherapists tell us that an unfriendly or absent father hurts a boy's emotional development, indirectly, as do marital conflicts. Effects of such paternal deprivations on young daughters have received less attention.

So, in the neonatal period, poor parental care constitutes stress for the baby due both to its unmet physical needs, and its emotional deprivation. Stress in very young lab animals may, indeed, simply result from hunger and thirst, but also from cold. Not being enclosed in the nest under the mother, surrounded by the rest of the litter, causes a serious fall in body temperature. Among human babies, also, hunger and thirst will cause them to cry, but also the discomfort of dirty diapers and gas pains may set them off. Too much of this and, for some babies, there may be long-term behavioral consequences.

But the long-term effects of stress do not just begin at birth. They can begin early in pregnancy. Moreover, males and females differ in the consequent deleterious effects. Tracy Bale and her student, Bridget Mueller, at the University of Pennsylvania, found that male but not female offspring exposed to stress early in their mothers' pregnancies could not respond adaptively to stress when they grew up, and could not enjoy things that laboratory mice enjoy, like sweet solutions. Part of the trouble with these little guys may have started with the placenta, a complex link carrying blood from the mother's tissues to the babies. Placental connections to males, but not female babies, were affected in their detailed biochemisty by stress. And the DNA in the genes expressing stress-related chemicals had transcription-interfering methyl groups added differently following stress. In Bales' words, "sex-specific programming (of the brain) begins very early in pregnancy" and the fetus, especially the male fetus, is vulnerable to perturbations of the mother. Indeed, Bale has found a rare time that males, on the average, are more susceptible to trauma than females, which could feed into the explanation of male-predominant autism. And other scientists' work agrees with Bales. They also find that among male offspring, there are profound effects of lousy mothering on the neurochemicals and hormones underlying stress responses. Young male victims of maternal separation show greater fearfulness in novel situations.

Another thing about early stress of the mother during pregnancy—following stress, the steroid hormone progesterone can be secreted from the adrenal glands. This is important, because progesterone can act as an anti-androgen—it can get in the way of the actions of androgenic hormones such as testosterone. Therefore, stress, through progesterone secretion, can inhibit the kinds of sex-differentiating actions of testosterone in the brain that have been discussed above in so many ways. What does this kind of experimentation in a laboratory animal have to do with early life events in higher species?

Understood, much of our knowledge about the consequences of difficulties in care for babies comes from work with simple laboratory animals. We can do such detailed biophysical work and neurochemical work in standard laboratory animals, such as rats and mice, that we sometimes lose sight of the greater complexities of the development of brain and behavior in higher species. So let's consider baboons. The dominance rank of a mother within a social group affects behavioral traits in her offspring. Notably, for our present discussion, the lowest ranking mothers raised sons who had chronically higher stress hormone levels than the sons who had been raised by high ranking mothers. Thus, from the get-go, the reactions to stressful experience by a young male baboon will likely be affected by maternal factors outside his control.

If that social factor has long-lasting effects in male baboons, an entirely different story has arisen over the last two decades with respect to the social effects of poor treatment of baby female laboratory animals. Michael Meaney, a world-renowned neuroendocrinologist at McGill University in Montreal, discovered that the stress of separations from the mother during the neonatal period had long-lasting consequences for the female offspring, especially for their eventual performance as mothers themselves. Lousy mothers give rise to females who themselves will make lousy mothers. And, even more striking—the female babies that were tended by excellent, attentive mothers, with lots of tactile contact between mother and baby, themselves grew up to be excellent, attentive mothers when they were adults. So, as they say, "Choose your parents well!"

Studying boys, you can see real anger within the first six months of life, and neuroscientific experiments are beginning to piece together how it comes about. In experiments with hamsters, males that had been traumatized early in life engaged in social behaviors with a degree of aggression unparalleled in observations with normal control animals. In parallel, we know that boys who were victims of early abuse or neglect are more likely to act with violence as teenagers. Both laboratory scientists and clinicians suspect that something is wrong with serotonin neurotransmitters in the abused animals and children. Serotonin usually restrains aggressive impulses. If, as suspected, the efficiency of serotonin neurotransmission is low in the abused individuals, then higher levels of aggression would result.

Even the descendents of Harry Harlow's monkeys fit this formula. Monkeys who have low levels of serotonin, and who are also deprived of interactions with their mothers, show unpredictable, extreme aggression.

Thus, the quality of maternal care will affect the nature of the offspring's social and emotional behaviors. Better care, better behaviors in a biologically adaptive sense. But there is more. Neuroscientist Frances Champagne and her team at Columbia University found that the *length* of maternal care by laboratory animals also makes a difference in the baby's subsequent behavior and neurochemistry. Longer care fostered higher levels of social interactions after weaning. Further, females who

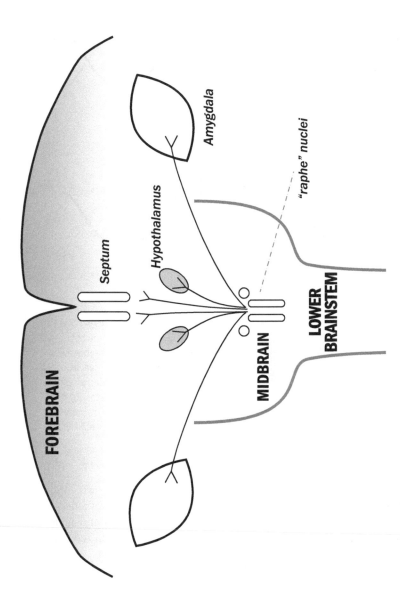

Projections of axons from serotonin-producing nerve cells in the midbrain toward the forebrain help to reduce aggression. The serotonergic nerve cell groups in the midbrain are called the "raphe" nuclei because *raphe* comes from the Greek word for fence, and these nuclei looked to classical neuroanatomists like a fence between the left and right sides of the midbrain. Important targets for these serotonin-bearing axons are in the hypothalamus, the amygdala and the septum.

The top drawing shows double-stranded DNA unwound and ready to be transcribed into messenger RNA. The DNA in the bottom drawing is also unwound, but transcription will be blocked. Here's why. The enzyme that makes messenger RNA (that subsequently will be used to make its corresponding protein) works by crawling along the DNA from the so-called 5' end of the gene toward the 3' end. Methyl groups (each composed of just one carbon atom and three hydrogen atoms)—always sticking up from the C (cytosine) of CG (guanine) pairing—will bring the passage of the enzyme to a halt. As a result: no messenger RNA, and no corresponding protein, perhaps for rest of the animal's or human's life.

were given longer care did, themselves, when raising their own litters, wean their pups later. These behavioral changes are accompanied by neurochemical changes that are different between females and males. Longer-raised females had higher levels of oxytocin binding in the hypothalamus, whereas longer-raised males had less binding. Since oxytocin secretion in the brain promotes higher levels of social interaction, it could participate in causing longer-raised females to stay, subsequently, with their own pups longer.

Transcription factors and coactivators

Inactive DNA is wound around little barrels of proteins called nucleosomes. Unwound DNA is more open proteins that want to bind to DNA and cause messenger RNA synthesis, "transcription factors," and their helper proteins, "co-activators." A closer look at one of those little barrels is given in the next figure.

How do the environmental actions—the effects of lousy treatment around the time of birth on behavior and on oxytocin secretion—last so long? We are not talking about genetic changes. Instead, we have a new word to explain long-lasting influences on behavior: *epigenetic*. Epigenetic changes refer to alterations in the regulation of gene expression, *without* changing the DNA nucleotide sequence itself. Most exciting, we are beginning to figure out the exact chemical changes that have these effects. One of the chemical changes involved decorates specific DNA nucleotides with a side group called a *methyl group*. Sticking this methyl group onto a gene promoter can make it vastly harder to turn that gene on. For example, molecular biologist Moishe Szyf, at McGill University School of Medicine, has evidence that methyation of the gene that codes for a stress hormone receptor can be increased by certain neonatal experiences. Because of this DNA methylation of a stress hormone receptor gene, the receptor's gene is about 40% less active. In the words of Columbia University professor Frances Champagne, poor expression of this stress hormone receptor gene, called GR, causes poor negative feedback control over the release of stress hormones, and therefore leads to higher chronic stress hormone levels.

The next figure shows a cartoon of this process. But these epigenetic changes are not limited to the DNA itself. Access to DNA by the factors in the nerve cell nucleus that initiate gene transcription is controlled by proteins called *histones*. Chemical changes of histones (illustrated by the figures on pages 157, 159, and 160) can turn totally inaccessible DNA into DNA which is open for gene transcription to be turned on. Either of these types of chemical changes—methylation applied to a gene, or histone chemical modifications where they cover up specific genes—can permanently alter expression of genes in the brain related to emotional and social behaviors. The stress hormone receptor gene targeted by Szyf and Champagne gives one fine example, but alterations in the synthesis or release of neurotransmitters that regulate emotional behaviors—serotonin, for example—will also be important to study.

Thus, as a consequence of this suite of epigenetic changes, the genes expressing neurochemicals and hormones that influence stress responses will be transcribed differently, and those differences in turn may produce female/male differences. For example, a neuropeptide called *urocortin 2* regulates circadian rhythms of stress hormones by decreasing them in females but not males. In turn, stress hormones combat inflammatory responses by the immune system, leading to immune differences between

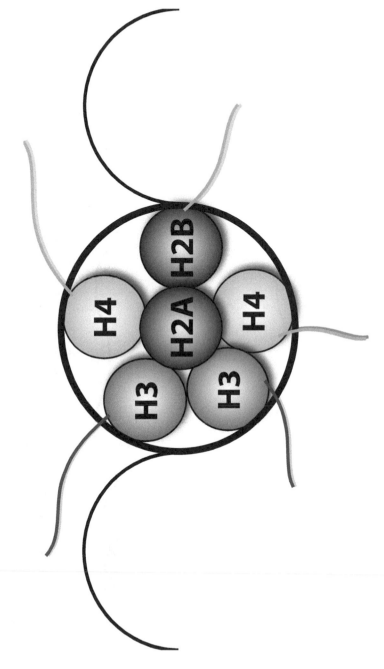

H stands for histone. A histone is a basic (meaning high pH, opposite of acidic = low pH) protein. Particular combinations of histones form the nucleosomes sketched in the previous figure. The tails of histones can be chemically modified in a variety of specifically regulated ways to form what C. David Allis, a molecular chemistry professor at Rockefeller University, has called a "histone bar code." When those histone proteinaceous tails sticking out of the nucleosome are chemically modified in just the right way, the DNA associated with them may become "open for business," as in the next figure.

Once the DNA of particular genes has been rendered open for business, transcription factor proteins can use their DNA binding domains (DBD) to fasten onto specific sequences of nucleotide bases (nucleotide bases in DNA A = adenine; T = thymidine; C = cytosine; and G = guanine. n = any nucleotide base at all; it's in there as an obligatory spacer to let proteins bind to DNA in the right way). Notice that two different transcription factors, estrogen receptor (ER, will have been loaded up with estrogens) and glucocorticoid receptor (GR, will have been loaded up with cortisol) fasten onto different sequences of DNA nucleotide bases. As a result, the transcription of estrogen-sensitive genes and cortisol-sensitive genes, respectively, can be started.

female and male that may last a lifetime. Esther Sternberg, a medical doctor at the National Institute of Mental Health, points out that sex differences in the dynamics of these stress hormones (and the brain's responses to stress hormones) might explain the much greater incidence of autoimmune diseases in females compared to males. In turn, some of these autoimmune diseases have implications for the occurrence of behavioral syndromes, such as the fatigue and pain syndromes suffered by females that I discussed in Chapter 8.

In laboratory animals, we do controlled, careful experiments that cause measured changes in behavior due to neonatal stress. In humans, the consequences can be more serious. People who had been abused as children and eventually committed suicide were shown, during postmortem studies, to have stress hormone receptor genes that were significantly less active compared to controls; namely, people who had died but had not been abused as children. Many of us feel that the epigenetic changes pictured above, applied to particular genes in specific parts of the brain, cause some of these unfortunate changes in animal and human behavior.

Puberty

The years between childhood and adulthood offer a huge opportunity to understand the relationships between changes in the body and changes in behavior. Although I am most interested in how the brain controls behavior, medical doctors who study puberty measure its stages in terms of bone age and pubic hair. Looked at medically, these measures, the hormonal changes, and the other biological changes of puberty, last much longer than you would think—as long as 9 years. On the average, girls enter puberty earlier than boys, by about 2 years, but there are wide variations among individuals. Thus, according to clinical endocrinologist Melvin Grumbach at the University of San Francisco, the most modern medical standards for the onset of pubertal development in the United States would set the earliest girls at 7 years of age and the latest at 13 years.

Exactly when a girl enters puberty depends on her environment. Grumbach and others have charted differences among countries and, in the United States, differences between girls with different racial backgrounds. The most obvious environmental factor that determines the

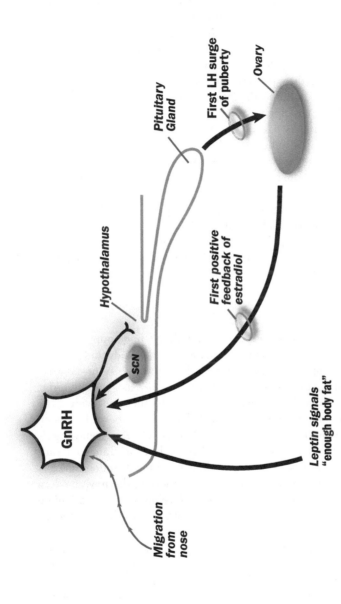

In the brains of females ready to enter puberty, the GnRH neurons will be in place, just in front of the hypothalamus, having migrated during brain development from the odor sensing surfaces of the nose. Interestingly, one of the same gene products that permits this migration is also, according to recent data from our lab, involved in regulating sex behavior. At the beginning of puberty, the GnRH neuron's response to estradiol kicks into "positive feedback mode"—estradiol causes more GnRH causes more LH (luteinizing hormone) causes more estradiol from the ovary, etc. That positive feedback dynamic can yield the first real surge of LH, leading to the first ovulation. However, signals from the body telling the brain that there is enough body fat are also required for puberty to begin. Leptin, a hormone secreted from fat, provides one important signal. The presence of adequate fat is important because it signals the brain that the female has good enough nutrition to feed babies.

time of puberty is the availability of food. Better nutrition, earlier puberty.

Environmental causes of behavioral change during puberty may work through hormonal mediators. While estrogens and growth hormones play important roles in kicking off the bodily changes of puberty, thyroid hormones and various growth factors also play permissive roles. Keep in mind that changes in stature and body shape can cause behavioral changes indirectly. A boy may act aggressively much more easily if he is bigger and more muscular than the other boys. His height depends on growth hormone. His muscularity depends on testosterone, among other things. And a girl may act flirtatious more readily if her body has begun to assume an adult female's shape, thus drawing boys' attentions.

Eventually, these environmental and hormonal causes must, so to speak, "reach in" to meet those exact neurons whose activity tells the pituitary to kick off pubertal hormonal changes. These GnRH neurons will have long since finished their migration from the nose into the hypothalamus, and at the beginning of puberty are waiting there to swing into action (think back to Chapter 2). Sergio Ojeda at the Oregon National Primate Research Center, and Ei Terasawa at the University of Wisconsin, have studied and summarized the most up-to-date information about how influences on these GnRH neurons cause the beginning of puberty. Some of their conclusions would not surprise us. As mentioned, the ancient neurotransmitter glutamate, which excites neurons, excites GnRH neurons and fosters the GnRH output that triggers puberty. The ancient transmitter GABA that inhibits neurons, inhibits GnRH and retards puberty. Signals from fat, informing the brain that nutrition is adequate, work indirectly to excite GnRH neurons and foster puberty.

However, Sergio Ojeda also discovered something else that surprises us very much. Not just nerve cells, but also their supporting cells called *glial cells* play an important role. Earlier, I talked about how these glial cells cover a large part of the surface of any GnRH neuron and, in fact, communicate with GnRH neurons by sending molecules through tiny holes in their membranes that match up with tiny holes in GnRH neuronal membranes, like the Palestinian Hamas' tunnels. Ojeda and his team found that growth factors and other chemicals produced in these glial cells act by traveling through these tunnels, or by receptors on the GnRH neuronal surface, to facilitate GnRH release and thus to facilitate the beginning of puberty.

All of these environmental and biological changes will mean a lot for boys' and girls' behavior. To quote endocrinologists Dennis Styne and Melvin Grumbach, "puberty is the biological process of attaining reproductive maturity, whereas adolescence refers to the psychosocial changes of the same period in our lives." So many sources of stress impact these adolescents that it would be impossible for me to name them all. For example, they are changing from the relatively sheltered, single-classroom environment of grade school to the larger, multi-teacher, confusing environment of junior high school or high school. They are unfamiliar with their changing body shapes. They cannot know exactly what is expected of them by their social groups, their parents and their teachers—and those expectations are not always the same, anyway. And, of course, there are sexual demands. Even though boys reach puberty later than girls, on the average they copulate sooner. Exactly how soon depends on parental influences, peer pressure, and the competing environmental demands on their time.

Jacqueline Eccles, social psychologist at the University of Michigan, tells us that adolescence is the worst time to impose any unnecessary stress on a girl. Harking back to Chapter 8, stress responses in young women are likely to be larger and last longer than in young men. This is because estrogens heighten stress hormone responses. In fact, molecular endocrinologist James Herman and his team at the University of Cincinnati discovered that estrogens do this by acting not only on the brain, but also on the adrenal gland, the major source of stress hormones. The consequences are severe for cognitive and emotional functions. Tracy Shors, behavioral neuroscientist at Rutgers University, reported that even as a stressful experience can actually help a form of learning in male rats, the same experience impairs that learning process in female rats. This finding may have implications for adolescent girls, as well. Further, Eccles said that because stress during adolescence will much more likely cause the girl to become depressed, parents should try to avoid social stresses such as changing schools and forcing their daughters to break into cliques already formed. Worst of all, Elizabeth Young and Margaret Altemus have suggested that "the onset of reproductive hormonal changes modulating stress systems at puberty may sensitize girls to stressful life events...." A real vicious cycle!

Before puberty, boys and girls have an equal frequency of depression and that frequency is low. Just about the time girls' breasts enlarge, the

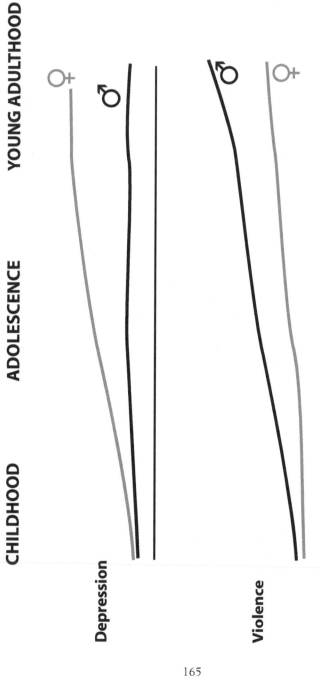

CHILDHOOD   ADOLESCENCE   YOUNG ADULTHOOD

Depression

♀
♂

Violence

♂
♀

Adolescence is the time period when sex differences in some of the most unfortunate aspects of mental life begin to become obvious. While young men are more likely to start becoming violent, young women are more likely to become depressed, especially if exposed to unusually stressful circumstances.

frequency of depression begins to rise suddenly. Bodily changes may lead to a negative self-image. Becoming a teenage mother is likely to make things worse, with young motherhood clearly associated with increased risk for female depression. Other exacerbating factors include loneliness and unemployment among teens. All of these circumstances help to account for the gender difference in the incidence and length of depressive episodes that arise among girls during puberty.

Positive peer relationships and positive parental and school influences that make for an easy adjustment to these changes will militate against depression. Higher education reduces the female/male difference in depression, as well.

For adolescent boys, rates of depression don't rise. They have other problems. Roughly in association with rising testosterone levels, they are more likely to become aggressive. Such behavior can range from simple strutting and posturing and self-aggrandizement, to violent, unmanageable aggression. Again, their social and physical environments play important roles. If the "cool" thing to do in a boy's neighborhood is to join a gang, in the absence of countervailing social influences, he will join that gang, and then he may be on his way to an aggressive and maybe even a criminal pattern of behavior. And, if he lives in terrible physical circumstances, a neighborhood in which it is impossible for him to imagine a positive social role as a young man, aggressive or violent behavior may result.

Some of these environmental effects on boys may be mediated by hormonal actions. A novel or stressful environment may be correlated with higher levels of stress hormones and lower testosterone levels. Under such circumstances, a boy may show signs of stress, but less aggression. It appears that in some adolescent boys, social anxiety and withdrawal militate against aggression, although, of course, they themselves are not desirable behaviors.

Not all of the problems with adolescent boys can be blamed on hormones. When the social support systems fail, we have trouble on our hands. And some problems with teenage boys derive from social traumas that impacted these boys much earlier in their lives. Consider boys who don't relate normally to others, even to the extent that they would be considered autistic. Long before their adolescent years, autistic boys not only will have failed to develop social recognition and bonds with others, but they also lack the social perspective and empathy that would allow them to act in a friendly and altruistic fashion. If autistic boys provide an example of things going wrong on the part of the person emitting a social behavior, what about when these boys are on the receiving end? Several authors have reported evidence that when we have been excluded from an emotionally important relationship, the social affront to us activates neural circuits involved in the control of our guts and in the appreciation of pain. Conversely, a supportive social environment can reduce the emotional impact of pain, and such amelioration depends on the health and activity of neurons in a part of the frontal cerebral cortex called the *anterior cingulate cortex*. Thus, even primitive biological events related to sex and pain are watched and regulated by sophisticated parts of our forebrains.

The presence of females, and the need to compete for females and other resources, complicate the situation for many teenage boys. First, the simple presence of a woman can increase testosterone levels in a young man, especially if he is feeling aggressive or dominant. Further, in a competitive situation, if a young man feels he is making a contribution to the outcome of the situation, his testosterone level will rise as a result, whether or not he wins the day. As Cheryl Sisk and her coworkers at Michigan State University showed, testosterone flowing in the blood, precisely during pubertal years, will foster agonistic behaviors and will be necessary for later effects of testosterone on aggressive, territorial behaviors. Paraphrasing Sisk—"testicular hormones during puberty organize neural circuits" that underlie territorial behaviors in experimental animals.

Despite our best efforts, we know that for some adolescent boys, in some social situations, things will get out of control—they will become violent, join gangs, commit crimes. What do we do about it? Of course, raising them in a clean, safe civilized environment helps a lot. That is, as opposed to environments that are physically and socially disrupted, a la James Q. Wilson's "broken windows" theory of crime, I assume that raising preadolescent and adolescent boys in a calm and well-ordered social setting will reduce antisocial behavior. Beyond that, we know that trying to avoid situations that humiliate the boy, and that preventing anonymity and social isolation by teaching him in a small school, have been shown to reduce violent behavior. Making sure that he has such strong social support that he is not tempted to join a gang will head off a lot of trouble. Avoiding alcohol consumption should ensure that minor acts of unfriendliness do not turn into impulsive acts of violence. Most of all, for adolescent boys, providing rites of passage—initiation rites, confirmations, and other religious recognitions of his impending manhood that offer positive visions of the boy's useful roles in society—will prevent the despair that leads to a downward trend in his social behavior.

All of these strategic warnings about boys' antisocial behaviors and strategies for avoiding their development remind us of the exactly opposite types of brain mechanisms discussed in Chapter 7, those for prosocial behaviors. Do these social and economic "environmental strategies" get any help from boys' natural neurobiological tendencies? Yes! We begin to learn how to form acceptable social relations early in life. In an earlier generation, the Swiss psychologist Jean Piaget watched how children naturally learned from each other how to play fairly. Paraphrasing Piaget, younger children learned from older ones, and the rules helped both because they facilitated social relations. And those who learn, survive. James Q. Wilson, a social scientist at UCLA, sets cooperative, friendly human behaviors in the evolutionary framework established by Charles Darwin. Natural selection determines features of the social behaviors of individuals who will survive to the age when they can reproduce, and then who actually will mate and have offspring.

The evolutionary biologists Robert Trivers and William Hamilton added a key insight. Being related to other boys in one's social group helps. In some cases, an individual can increase the degree of prolongation of his/her genes into future time by helping others who have some of those genes, although not all of them. The name of this idea is the theory of "inclusive fitness for reproduction." A person's fitness included not only his/her own ability to produce babies, but also included the ability of his/her kin who have some common genes, the more the better. As a result of this process, boys will have inherited the tendency to behave reasonably, especially towards one's relatives. The sociologist James Q. Wilson not only employs this thinking to understand why we exhibit group-oriented, positive, friendly behaviors, but also recognizes that whenever people treat each other in a fair, sympathetic manner, they are exhibiting an essential, underlying understanding of the importance of reciprocity. Wilson would agree with my argument in *Neuroscience of Fair Play.* He says: "The norm of reciprocity is universal." If we do a favor, we expect one in return. If we receive a favor we cannot return, we are distressed. Wilson stretches his argument from the mechanisms for parental care, to a desire for attachment and affiliation—the desire for a friendly social environment enjoyed by the individual and fostered by his/her own ethical, fair, sympathetic behavior.

Again, in terms of natural tendencies that help boys and girls to bridge the difficult adolescent years, even as scientists we must begin to think about love. Love yanks us out of our self-centered cocoons in order to allow a consideration of another person's interest. Even when contemplating a human feeling like love, we must recognize that altruistic love springs from evolutionary strategies for survival, and derives from the formation of the parent–infant bond. Putting these thoughts together with those in Chapter 7, I would argue for a continuum of brain mechanisms for mating and parental behaviors, through normal human cooperation, through love.

Thus, we are reminded that the adolescent years are critical years for the development of behavioral tendencies that essentially will distinguish the emotional and social dispositions of men and women. Despite all the things that can go disastrously wrong, we really are talking about countervailing forces that come into play during these years. For boys joining gangs, aggression and other antisocial acts could erupt, *but* as I have argued in *Neuroscience of Fair Play,* our nervous systems are wired for kind, reciprocally beneficial behavior, so that if the boy is given a chance, he'll come out of adolescence alright. Girls may be liable to depression, *but* offering the right kind of social support, and cleverly avoiding unnecessary stress, can get them past these hurdles and safely into their adulthood.

The Story So Far

If these few words have treated some of the social woes that, separately, can afflict girls and boys during critical developmental periods, their impacts may seem mild compared to the biological disorders of sexual development discussed in the next chapter.

TEN

"Sex Gone Wrong"

Normal sexual behaviors depend on large numbers of delicately balanced mechanisms—anatomical, hormonal and psychological—to "work right." Precisely because of the precision required of many biological and psychological processes for normal reproductive development, there are many ways for things to get thrown off. This chapter will deal just with a few of the many excruciating problems of sexual development and performance in children and young adults. The most obvious problems derive from anatomical defects. Developmental abnormalities in females or in males can force agonizing choices, medical and psychological, for the patients' families. Other problems are hormonal. Hormone feedback problems can cause developing females to be exposed to high levels of androgens, and in males, androgen receptor problems can interfere with normal actions of testosterone. Finally, some unfortunate situations seem to have arisen purely psychologically, and can even sink to a level of criminality.

Nature's Anatomical Mistakes

When Justine Schober, M.D., enters her operating room about seven o'clock in the morning, the patient she encounters may weigh only ten pounds and measure about twenty inches. Dr. Schober is a urologic surgeon who specializes in fixing the genital problems that some tiny babies were born with. Some of these problems have to do with abnormalities of sexual development.

All sex behaviors, male and female, depend, at the end, on our sexual apparatus. Some infants, in their different ways, have certain anatomical problems that usually can be remedied surgically. Other problems derive from abnormalities of hormone action, for which sophisticated nonsurgical approaches are used. These various abnormalities ultimately relate to functions of gender identity, sexual behavior and function, and subsequent reproduction, which surgical, hormonal or psychological treatments are intended to resolve.

To the pediatric surgeon, ambiguous genitalia are often apparent immediately at birth. "Is this the anatomy of a little boy or a little girl?" Once the parents and the medical team have consulted and reached a consensus, extremely difficult surgery may be chosen.

How do the medical team and the parents decide upon a course of action? Should the child's physical "gender assignment" be male or female? Dr. Schober and many of her surgical colleagues try to render the baby's genital anatomy and psychological sex assignment in concert with the child's chromosomal sex most of the time. But because the patient is so small, the outcome, if surgery is chosen, will not be immediately known. The child will be followed by parental observation with psychological support, and by medical evaluation. Each will have their own opinion of the child's behavioral outcome.

In some cases, a genetic defect will have led to anatomical changes that obscure normal female genital development. In the female, therefore, a type of surgery called "feminizing genitoplasty" may be in order. This could take the form of (1) cliteroplasty, during which portions of the visible, external clitoris and portions of the internal clitoris may be reduced in size or removed; (2) labioplasty, during which the foreskin or the phallic shaft skin is lifted as a flap, split, refolded, and angled to create the labia majora and labia minora; or (3) vaginoplasty, during which, in the simplest case, an incision must be made through the fused

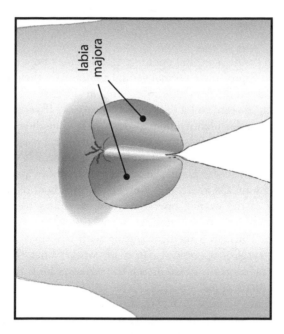

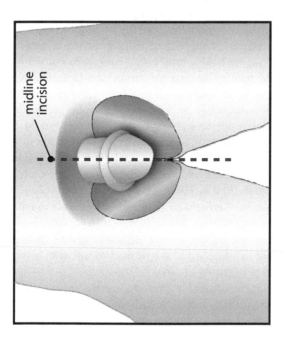

If a female baby is born with genitalia that look like a male's, a pediatric urologic surgeon can make a midline incision, as shown in this simplified sketch, split and refold the skin, and achieve a labia majora.

labia in order to create a vaginal opening. In all types of operations, the surgeon hopes to approximate normal genital anatomy for the purposes of sex behavior and eventual motherhood. And in all cases, surgeries at much younger ages seem to be easier on the patient. If genital appearance reinforces gender identity, one might expect that there would be better outcomes, particularly with respect to achieving a psychologically stable gender identity that enables an individual comfortably to function sexually as a woman. For example, in studies reviewed by pediatricians Albert Ong and John Gearhart, at Johns Hopkins School of Medicine, more than 80% of the women operated upon as infants were sexually active, and about two-thirds reported having orgasms. Their gender identity was female. Thus, in a medical setting with a serious anatomical problem, these surgical operations can really work. What Dr. Schober and other urologic surgeons frown on are the requests for surgery on the genitalia for cosmetic or other purposes.

When it comes to males, Dr. Schober finds that she must operate on as many as a hundred patients per year. Some infant boys may have malformations of the penis, such as hypospadias or a micropenis. In hypospadias patients the urinary opening is misplaced, often severely enough as to look like a vagina. Because the penis may be severely curved, such cases may require surgery that will permit sexual intercourse as a male. Boys growing up having undergone surgery to correct hypospadias have been found to have normal psychosocial adaptation, regardless of the severity of the initial anatomical problem. Gender-typical masculine behavior can develop normally, but if the number of hospitalizations skyrockets, school problems and emotional problems can arise.

Regarding micropenis, even though it is possible for such patients, having grown up without surgery for such conditions, to have sexual intercourse, such men usually do not find regular partners, and even fewer have sperm in their ejaculate. They assume masculine gender role identities and can have erections. They are comparable to typical men in regard to gender issues, body image, social fitness, sexuality, and family adjustment. They are, however, often dissatisfied with their penile size, which, according to some studies, can lead to a fear of rejection and a lower likelihood of pursuing a sexual relationship.

In such cases, the doctor has two options that are not mutually exclusive. One is to treat the penile skin with an androgenic hormone cream, like one containing dihydrotestosterone. The penis will grow rapidly, and

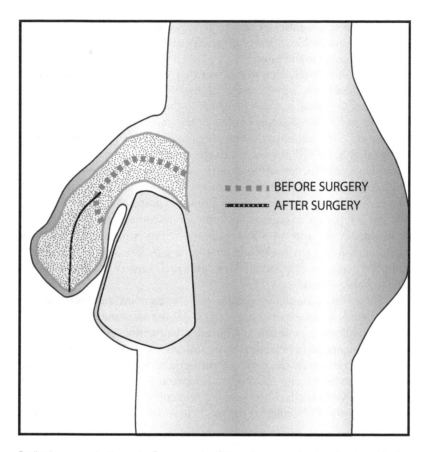

BEFORE SURGERY
AFTER SURGERY

Pediatric surgery in the male. For example, if the urinary opening has developed in the wrong direction, it can be redirected (darker black line) toward the tip of the glans penis, to achieve normal masculine sexual function.

under the right conditions can reach a substantial fraction of normal adult length. The other option is surgery. In the words of CRJ Woodhouse, a urologist at University College London, the "surgical enlargement of the penis is limited by the inability to make erectile tissue." Other, non-erectile parts of the penis can be reconstructed with skin flaps. However, Woodhouse concludes that "as a general rule, a small penis should be used as the basis for sexual function. Every effort should be made to help men to have sexual satisfaction and to this end, a knowledgeable sexual therapist is invaluable. Surgery on the penis is often poor treatment for problems that lie in the brain."

Woodhouse is referring to the fact that, contrary to popular opinion, the penis does not "have a mind of its own." Its erections are under the control of a small group of neurons in the lower spinal cord. They, in turn, are under the inhibitory control of the lower brainstem reticular formation, reticular neurons that are regulated by the hypothalamus and many other nerve cell groups. So, in the marginal case of micropenis, for example, successful sexual intercourse and holding a mate may depend quite as much on neural and psychological factors as on the condition of the penis. Nevertheless, following penile hypospadias surgery, marked success is reported, as regards sex behavior outcomes. About 90% of men were sexually active and 100% were potent. Surgery for micropenis is not as successful. Surgery for penile lengthening is done only in adulthood, after the penis has stopped growing, and such surgery can be described as unreliable at best.

These decisions about gender can be complex and excruciating. In the words of an expert on pediatric gender assignment, Heino F.L. Meyer-Bahlburg, a professor at Columbia University, "clinical policies of gender assignment in newborns with ambiguous genitalia are dependent on (1) the clinicians' theoretical assumptions concerning the determinants of gender, (2) the relative importance attached to outcomes such as gender dysphoria, fertility, sexual functioning, sexual orientation and general quality of life, and (3) the medical treatment options available at the time."

Parents raising children with ambiguous genitalia have sometimes had to "choose sides" between two warring camps of medical opinion. On the one side were the adherents of the late sexologist John Money, at Johns Hopkins School of Medicine. He maintained that the sex assigned to a child, the so-called "sex of rearing," would dominate that child's gender-related behavior and that the child would be perfectly happy with such a sex assignment. On the other side were biological and genetic experts, who said that chromosomal sex with consequent sex-differentiated hormone levels should determine how the child is raised. For example, a patient who, chromosomally, is an XY male but with androgenic hormone deficiency, raised as a female, would happily change from female to male when at puberty the full medical picture was seen. If you were to look across all the cases reported during the last fifty years, you could find patients whose development would make one side look bad, and other patients whose development would make the other side look bad. I hold with the opinion of Heino Meyer-Bahlberg, who says that families should be consulting with a team of experts that cover the range of medical, genetic, psychological and social issues that surround any decision of gender assignment. In the words of clinical endocrinologists Y-S Zhu and J. Imperato-McGinley, "in normal males, the sex of rearing and the androgen imprinting of the brain act in concert to determine the expression of the male gender." Environmental and hormonal factors work together to determine gender identification in the ideal case. But because of the complexity of the medical and psychological issues, and the number of different factors bearing on the decisions to be made, results are not always comfortable for the patient or the family.

Ambiguous genitalia and gender identity cause a lot of stress for the parents. In the words of Ieuan Hughes, professor of pediatrics at the University of Cambridge, "if a family is already vulnerable to psychological distress, educational sessions (about the child) alone may not suffice. Parents who are experiencing escalating distress may compound the problem for the child." Psychologists sometimes need to train parents "in the skills that they need to raise the formerly genitally disordered child in a healthy and emotionally successful manner."

Hormone Actions

Consider the young patient, apparently a female, who has not begun to have menstrual cycles. She has ambiguous genitalia, and features a large clitoris. The urinary and genital tracts have fused, the way they would in a male. Yet her sex chromosomes are XX, as in a female. How did this happen, and what can be done?

She might have a condition called congenital adrenal hyperplasia (CAH), in which a genetic change sharply reduces the activity of an enzyme, a biochemical that is essential for the production of stress hormones by the adrenal gland. Because of that, the brain and the pituitary gland keep on telling the adrenal gland to produce more hormones. There is no "negative feedback" to regulate adrenal gland production. There lies the problem. As well as trying to put out the stress hormones that it cannot produce, the adrenal gland also pumps out male-type androgenic

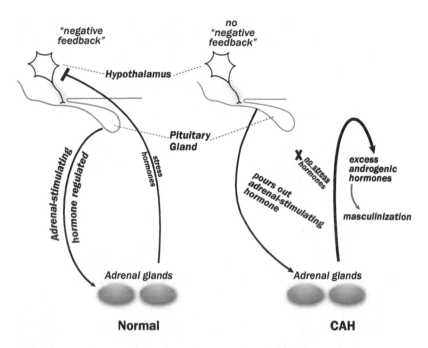

In a girl with congenital adrenal hyperplasia (CAH, right side of drawing), the adrenal gland cannot make the stress hormones that normally would enter the brain and cause "negative feedback" that, in turn, regulates adrenal gland output (Normal dynamics, left side of drawing). Instead, in CAH, the unregulated adrenal gland pumps out excess androgenic hormones (which are part of the same biochemical pathway of steroid hormones as are stress hormones), with the effect of masculinizing the CAH patient's physiology and appearance.

sex hormones. Because of the lack of regulation of adrenal gland hormone production, these male sex hormone levels go far higher than they are supposed to. As a result, they produce abnormalities in the very part of the lower body that is essential for sex behavior. Excess androgenic hormones cause severe enough anatomical changes in the genitalia to disrupt normal female sexual assignment, or even severe enough to cause some parents to raise the child as a boy.

If she had had recurrent surgeries during infancy to separate the urinary and genital tracts and to reconstruct female-type genitalia, the young patient might have avoided a lot of difficulties. However, these surgeries risk loss of sexually important sensation and, in addition, the vaginal opening may not be large enough for normal sexual intercourse. Another approach uses the administration of the stress hormones the baby lacks. Beginning prenatally, such a stress hormone is injected to her mother, so that the baby's adrenal glands do not overproduce male sex hormones.

CAH is the most common cause of genital masculinization in baby girls. The behavior of CAH patients is often masculinized as well. But children with the highest degrees of genital masculinization are not necessarily those with the highest degrees of behavioral masculinization, defined in terms of high levels of activities that traditionally were recognized as typical of males. Does a child play with dolls and concentrate on the insides of little play houses, or does the kid play with trucks and guns, and range into a much larger space? Does a child avoid strong physical contact with other kids, or engage in rough and tumbling patterns of play? How good is the patient's verbal fluency, compared to spatial comprehension, for example with mental rotations? Are there symptoms of autism, much more frequent in males? Could we call the patient "tender minded" (intuitive, sensitive) or "tough minded" (self-reliant, no-nonsense)? Is the kid really drawn to infants? Or, instead, when envisioning hypothetical situations, does the patient opt for attitudes of verbal or physical aggression?

In one recent and very important study, psychologist Melissa Hines and her team at the University of Cambridge found that CAH patients compared to unaffected female relatives were masculinized. They showed greater aggression and were more tough-minded rather than tender-minded. They were not as interested in infants as their female relatives.

For older patients, to whom is she sexually attracted? Psychologically, patients are more likely than other XX females to feel masculine. They tend to be sexually attracted to females. Their psychological change of gender from female to male is gradual, and extends into adulthood.

What about males? Failures of male sexual differentiation can also result from problems with hormone-producing enzymes, in this case androgenic hormone production, as well as androgenic hormone action. As Julianne Imperato-McGinley and Yuan-Shan Zhu, at Cornell University Medical College, have pointed out, genetic mutations leading to the loss of either of two enzymes essential for testosterone synthesis can lead to the kinds of ambiguous genitalia I discussed earlier. Breakthroughs in the genetic, the biochemical, and the psychological aspects of these conditions have emanated from the medical scientists intensely studying small populations in which such mutations are present in a large number of males. For example, Dr. Imperato-McGinley had the unique opportunity to follow affected members in the Dominican Republic over 25 years. Her patients, bearing X and Y sex chromosomes like normal males, had phalluses that were like clitorises, and had hypospadiases that were like vaginas. Their families very reasonably raised them as girls. Things became very confusing when, at puberty, they had increased muscle mass and a deepening of the voice. What happened then? In Dominican society, most of the adolescents preferred to assume a male gender role, but some chose to retain a female gender role. In such sexual assignment crises, the expert Heino Meyer-Balhberg would not allow a family to fend for itself, but would instead call for the kind of multidimensional medical and psychological intervention I ascribed to him earlier.

If one type of problem is the failure of production of androgenic hormones, another type of problem is the failure of response to androgenic hormones once they are produced. Androgen sensitivity depends on specialized proteins called androgen receptors (AR). Here is how it works. Testosterone (T) is a hormone that is easily soluble in fat. Therefore, it swims across the cell membrane—be it a genital area cell or a nerve cell—easily. After crossing the membrane, it encounters a protein whose three-dimensional shape harbors a binding site, a landing site, a dock for testosterone but not for other kinds of hormones. Once the T and the AR get together, the AR changes shape, and its protein signal for getting into the cell nucleus is revealed on its surface. When it has arrived at the cell nucleus, it sits on certain portions of the DNA (see Chapter 9) and changes gene expression. Like any other genes, those genes encoding androgen receptors (AR) can under go mutations that will interfere with their function.

So, this beautiful system of steroid hormone action can go wrong. Some males, bearing Y chromosomes as expected, are completely insensitive to their own androgenic hormones like testosterone. They look like females. They have breasts, vaginas (though short), and may not be discovered as genetic males until their lack of menstrual period is investigated. These individuals have testes and normal androgenic hormone secretion, but they can not respond to their testosterone for reasons like

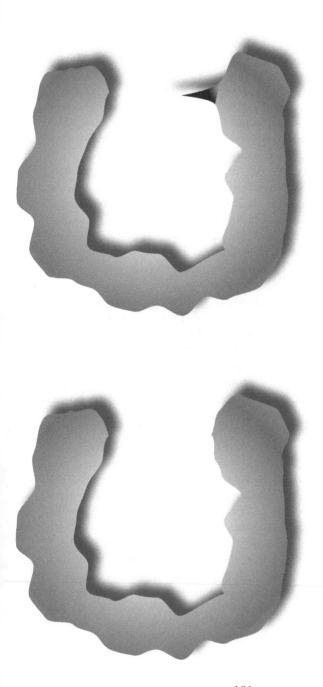

Hormone-Sensitive

Hormone-**IN**sensitive

On the left, a cartoon of the portion of an androgen receptor (AR) that is well formed to receive testosterone, the steroid hormone that would slip into its cavity and be bound there. From the top to the bottom of the cavity would measure a small portion of a millionth of an inch. On the right, a cartoon of the sort of result that can happen as a result of a mutation of the AR that would make the AR unable to bind testosterone, and so the tissue in which these cells with mutated AR are located would be unable to respond to testosterone. As sketched, even a small change in the size or shape of the cavity can block the testosterone from getting into the cavity, thus rendering the cell insensitive to testosterone.

181

that illustrated above. Because of mutations, the expression of the AR may be very low. Or, it could be expressed from the AR gene at normal levels, but could end up as a misshapen protein. For example, if the portion of AR that binds to DNA and changes gene read-out is covered up or altered, AR can not do its job. Most important for explaining a loss of masculine function, the tiny pocket in the AR protein that accepts and binds testosterone might be too small, and the testosterone can not get into it.

In terms of the genetic causes of complete androgen insensitivity, we have some clues. It is transmitted to the young boy through the mother, suggesting that the problem lies on the boy's X chromosome. Since the AR gene is located on the X chromosome, and since more than 200 different types of mutations of the AR gene have been described—ranging from subtle but serious ones illustrated in the previous figure, to a complete absence of the AR gene—we probably have the explanation of complete androgen insensitivity in hand.

As a result of androgen insensitivity, both of the external genitalia and of the developing brain, the patient will be raised as a girl. What happens at puberty, when menstrual cycles fail to appear? What to do?

In cases like this, Heino Meyer-Bahlburg distinguishes "true sex" from "optimal gender." "True sex" refers to a medical strategy dating from the nineteenth century that says: "assign the individual's sex according to that person's chromosomes, and everything else—gender identity, gender role, courtship, love and psychological health—will fall into line." The medical strategy Meyer-Bahlburg calls "optimal gender" requires that the gender assigned to the child shall carry the best prognosis for reproductive function, appearance, medical procedures, and the patient's overall happiness. Will genital surgery be required? Meyer-Bahlburg wants the family to consult with medical and psychological professionals to reach a consensus. And the team cannot operate by reflex. Yes, CAH females as a group are behaviorally masculinized but, in his words, "with much interindividual variability." Likewise, XY men who have not produced testosterone, or are insensitive to testosterone, are feminized, but with differences from case to case. Following partial loss of development of the male genitalia, as many as half of patients may choose a female to male gender change. But with complete androgen insensitivity, Meyer-Bahlburg reports that "no female to male gender change has been reported."

Quite a variety of situations can cause testosterone levels in the blood to be too low. Most of the ones I know about are genetic. For example, Klinefelter Syndrome hits men who have an extra X chromosome. Instead of XY, they have XXY. Their testes are small, they have testosterone concentrations that are below normal. Genetically engineered XXY mice and Klinefelter Syndrome patients both show sexual preferences for males. The patients show impaired male sexual identity and, overall, a decreased strength of sexuality.

Men whose testes simply do not produce enough testosterone, "hypogonadal" men, can easily be treated with testosterone. Injections of testosterone (T) every two weeks or so produce large flucuations of T in the blood and thus are not desirable. Tiny biodegradable spheres loaded with T that can be administered may turn out to be the best way to treat men whose gonads are not doing the job. As a result, they will have more energy, greater sexual desire, and a heightened mood.

One Form of Psychological Problem, A Criminal One

Finally, some problems with reproductive performance have to do entirely with abnormal sexual behavior, understood at the psychological level but treated hormonally. With respect to abnormal levels and types of sex behavior by males, consider the exact opposite end of the spectrum. These are men, criminals, who try to have sex with children, so-called "pedophiles." These deviants have to have their testosterone production turned off! In the Czech Republic they simply are castrated. Dr. Justine Schober, of the Hamot Medical Center, has used a more subtle but completely effective approach by making use of a peculiar characteristic of neuroendocrine systems in order to accomplish just that, the abolition of testosterone production. The cells in the pituitary gland (the "master gland") that tell the testes to produce and secrete testosterone, require the GnRH coming from the brain—from those GnRH cells whose migration I traced from the nose—to arrive in the form of tiny pulses, separated by many seconds when no GnRH arrives. If the GnRH arrives all the time—steadily—these pituitary gland cells are actually shut down. Dr. Schober made use of these cells' requirement for pulsatility by administering an artificial form of GnRH as a long-acting preparation. Thus, she shut down the pituitary gland cells, and the testes did not produce or

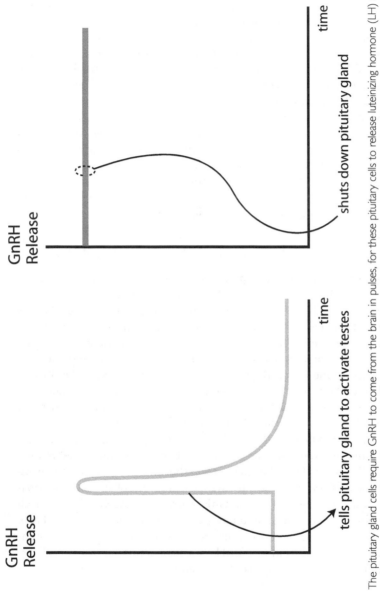

GnRH Release

tells pituitary gland to activate testes

time

GnRH Release

shuts down pituitary gland

time

The pituitary gland cells require GnRH to come from the brain in pulses, for these pituitary cells to release luteinizing hormone (LH) and activate the testes. So, the temporal pattern sketched here on the left works. The steady, high level on the right does not. Therefore, achieving a steady high level through medical administration of a GnRH-like chemical compound can shut down the testes of a sex offender and, as a result, greatly reduce his libidinous tendencies.

secrete testosterone. As a result, the sexual fantasies and illegal sexual behavior by these pedophiles were significantly reduced.

As we understand from the news all too frequently, identifying and treating men whose ranges of sexual desires reach a level of criminality has become a significant problem for medicine and public health.

Genes and Environment

How do all these problems come about? In too many cases we don't know. For many of the abnormalities of sexual development, we point to genetic causes. Chromosomal alterations, genetic abnormalities, play a huge role, and I am trying to give them their full due. But, I also worry about artificial chemical compounds in our modern industrialized environment. Some of these chemicals get into our bodies and act a bit like hormones to which they are chemically similar. Because they are not regulated, in their amounts or in their actions, in the exquisite manner that our own native hormones are regulated, and because they can cause untoward effects, they are called "endocrine disrupters"—they break up normal bodily patterns of hormonal actions.

Of direct relevance for this chapter, environmental chemicals that mimic estrogens, when applied to experimental female animals early in life, both masculinize those females and also reduce their feminine characteristics, anatomical and behavioral. Under circumstances where the concentrations of these endocrine disrupters are allowed to build up—for example, through the re-use of water in the Netherlands, or by very high consumption of certain fish by mothers in the United States— they are even more likely to cause abnormalities of sexual development. Many scientists also worry about another chemical, Bisphenol A, that is used in making plastics, baby bottles, the linings of tin cans, dental sealants (!), and, according to scientists Mary Ann Ottinger and Fred vom Saal, is one of the top 50 chemicals in production in the world. Bisphenol A binds to estrogen receptors in the body and thus has the capacity to increase the growth of estrogen-dependent tumors in the body, for example in the breast and in the uterus. For our present purposes, note that stimulating estrogen receptors early in brain development will disrupt normal female behavioral development, and increase masculine characteristics of behavior whatever the mammalian species, lab animals or humans.

Plastic is one thing, but what about dietary substances that are supposed to be good for us? Many vegetarians, including myself, eat a lot of soy as a source of protein that substitutes for meat. But soy contains chemicals called phytoestrogens, plant estrogens. They bind to one form of estrogen receptor, ER-beta. I am sure that ingestion of substantial amounts of phytoestrogens during development will affect brain and behavior, but the laboratory science remains to be done.

Reflection

When you look at the number of biochemical reactions needed to produce the normal complement of sex hormones in the female and male, and when you envision the anatomical job that must be completed properly to produce normal genitalia, you can end up wondering why the number of developmental problems is not greater. Scientists have proceeded quite efficiently in their analyses of the chemistry and the anatomy. However, in the case of any individual patient, once the genetic, the hormonal, and the anatomical facts are clear, people may disagree markedly about exactly what should be done. The field was held up for a long time by opinionated theorists espousing the extremes: "gender role and identity should be determined one hundred percent by chromosomal sex" versus "the way parents assign the child and raise the child will lead to perfectly fine gender role and identity." Now, Meyer-Bahlburg's emphasis on getting the parents together with the geneticist, the endocrinologist, the surgeon, and the psychologist in order to optimize the reproductive arc of development for that very child, will rationalize this field of medical practice.

The Story So Far

So much of the discussion in this and previous chapters has hewn closely to the medical and scientific literature that I have not always brought scientific developments home, in a manner that bears on our current human concerns. I will do so in the next, final chapter.

ELEVEN

Bottom Line

We have walked through the entire field of scientific work that analyzes the determinants of sex differences in behavior: Y chromosome, X chromosome inactivation, prenatal hormones, neonatal hormones, testosterone from the testis, androgens from the adrenal glands, prenatal stress, neonatal and pubertal environments, and so forth. What do all of these mechanisms mean for how we view our own gender roles—our assumption of behavior patterns, which society expects to be different between men and women—and the gender roles of those we love?

Two of the giants in the history of neuroendocrinology and behavioral endocrinology tell us what we can expect to infer from a book like this, and how we, as scientists or as citizens, should be cautious in our conclusions as we try to reason from scientific facts to thoughts about our own lives. Margaret McCarthy, whose work I described earlier, wrote an article called, "When is a sex difference *not* a sex difference?" (italics mine). She warns us that differences between adult males and adult females, in aspects of behavior that have nothing to do with reproduction, need not reflect underlying differences in brain structure but instead, may simply be due to whatever hormones those individuals have on board at the moment. Change the hormones, change the behaviors. Indeed, we need to think not only about the hormones themselves, but also where they are produced. New evidence suggests that the brain itself can produce small amounts of estrogens. Estrogens produced in the brain may have different effects on a woman's behavior, compared to estrogens produced in the ovaries. Androgenic hormones produced in the adrenal gland may have different effects on a man's behavior, compared to androgens produced in the testes. Furthermore, sometimes sex differences in measurements of learning have nothing to do with "ability" but instead have to do with sex differences in strategy—differences that, for example, serve females very well in some types of tasks and not very well in others.

The other authority, Michael Baum, former editor of the flagship journal *Hormones and Behavior*, wrote the article, "Mammalian animals of psychosexual differentiation: when is 'translation' to the human situation possible?" Clinical experience confirms that the large number of genetic and hormonal actions that I and others have gleaned from the literature on laboratory animal experiments do pertain to human behavior. However, in Baum's words, "results of animal studies can, at best, provide only indirect insights into the neuroendocrine determinants of human gender identity and role behaviors." He says this, in part, because of specific psychologic factors that separate sexual phenomena in humans from those in laboratory animals. For example, the greater dependence of human sexual attraction on visual signals distinguishes them from the strong effects of olfactory and pheromonal signaling in many lower animals. And the hormone-based extrapolations from studies of animal brain and behavior do not always work. For example, although the scientific literature has proven the importance of androgenic hormones during prenatal or neonatal critical periods of sexual differentiation of brain and behavior, *male-to-female* transsexuals did actually have the normal (not abnormally low) fetal androgen exposure, and *female-to-male* transsexuals had no excess of androgens. But Baum also had broader and more abstract considerations in mind. He knows the psychological importance of how a child of indeterminate sexuality is raised by his family, and the culture in which that family is operating. Baum distinctly feels that "the concepts of gender identity and gender role are uniquely human," and so it would be impossible to extrapolate directly and unreservedly from experiments on animal brain and behavior to resolve questions about human romance.

All of this having been said, some evidence does point to believable psychological differences between men and women that are of interest to the neuroscientist. The differences start with children playing. Recall from Chapter 10 the work of Melissa Hines, at the City University of London, who summarized the evidence that little boys have much different patterns of playing than little girls. Boys are more likely to choose trucks and guns to play with, to be outside, and to build stuff. Girls would rather be inside, on average, playing with kitchen things and cosmetics. Girls are more likely to be interested in babies. Boys are more likely to play in a rough-and-tumble fashion. Girls, not. These differences in play did not happen overnight. This is not a post-Industrial Revolution phenomenon. Kim Wallen, from Emory University, has charted sex differences in preadolescent monkeys, and found them pretty much the same as in humans. Juvenile male monkeys liked the wheeled toys, and indulged in rough-and-tumble play.

What about adults? The most convincing evidence, in my view, has to do with social relationships. Males are interested in hierarchies and in power. Oliver Schultheiss, and his colleagues at the department of psychology in Harvard University, not only measured power motivation in men, but even found that men high in power motivation, and only those men, had higher testosterone levels after imagining success in a contest of dominance. Further, their high testosterone levels continued only if they actually won.

In fact, the evolutionary psychologist David Buss, at the University of Texas, thinks that men's competitive, power-hungry motivation would even extend to their choices of women. In their study, men—much, much more frequently than women—would say that they chose their mate in order to gain status and reputation.

Some authors talk about personality differences between men and women. I must admit: the reports that, on average, women are more likely to seek agreement, exude warmth and show an "openness to feelings—whereas men demonstrate more assertiveness and an "openness to ideas"—sound to me to be very culture-bound. Likewise, Shelley Taylor's viewpoint, mentioned in Chapter 7, that women are more likely to empathize, to "tend and befriend" during difficult times, rather than fight. Authors who summarize the data from studies carried out years ago, about tender-mindedness in women, have to face the fact that currently, female corporate leaders, politicians, and soldiers break the pattern and reveal the considerable flexibility of modern societies with respect to social roles for able women. That is, culture plays a huge role in shaping the desire for power, the degree of competitiveness, and so forth. These personality differences are not written in stone! But that has not kept neuroscientists from looking for sex differences in the brain that give rise to these behaviors.

Brain Sex and Gender Identity

Indeed, there are anatomical differences between men's and women's brains, and some of them, theoretically, could contribute to the psychological differences I just discussed. Here is a warning, however, before I get into the details of sex differences in hypothalamic nerve cell groups involved in such functions as erection, ejaculation and aggression. Over the years, a considerable number of sex differences have been reported. But, in the words of UCLA professor Roger Gorski, "Brain sex is regional and specific, not global and general." What Gorski means to say is that sex differences in particular groups of nerve cells should not be misconstrued to indicate broadneuroanatomical differences of cosmic psychological significance. Therefore, I'll restrict myself to a couple of examples that have been proven to be reliable, and where the functional meaning seems to be understandable. For example, as discussed in Chapter 4, in the preoptic area of the brain there is a cell group called the *sexually dimorphic nucleus* that is more than twice as large in men as in women. Nerve cells in this group likely allow the man to ejaculate. And, just a little bit back toward the hypothalamus, Dick Swaab and his team at the Netherlands Brain Research Institute found another cell group, the central nucleus of the bed nucleus of the stria terminalis, that was two-and-a-half times larger in men than in women. These likely serve aggressive acts.

Swaab and his colleagues in Amsterdam also have reported differences that follow gender identity. They were provoked to make these observations by knowing that people often "have the feeling" of being male or female from childhood onwards, and thus might be influenced in their choice of gender identity by specific neuroanatomical features. Swaab's team measured numbers of specific types of neurons in a particular neuronal cell group in the brains of genetically male patients who

FEMALE

MALE

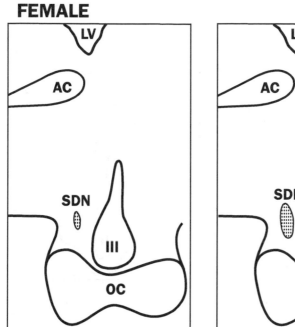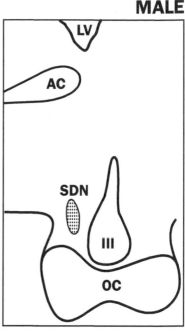

Neuroanatomists in the laboratory of Dick Swaab, former chief of the National Institute for Brain Research in Amsterdam, the Netherlands, found a sexually dimorphic nucleus (SDN) in the preoptic area of the human brain. It is about twice as large in a man's brain as in a woman's. The preoptic area is just above the optic chiasm (OC), and surrounds a "landlocked lake" of cerebrospinal fluid; the third ventricle (III) is below the anterior commissure (AC) and the lateral ventricle (LV).

were transsexual and changed their genetic identity to female. They found that numbers of these particular neurons—nerve cells that produce a small fragment of a protein called *somatostatin* (SOM)—were almost twice as large in males as in females. Amazingly, in male-to female transsexuals, Swaab's team found the female number. As an aside, I note that these cells are in the same group of neurons, the bed nucleus of the stria terminalis, that are tied to androgenic hormonal effects on aggressive behavior. Whether these differences in neuronal numbers in transsexuals are the causes of the gender identity switch, the results of it, or

FEMALE   **MALE**

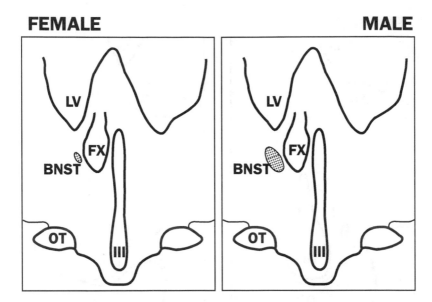

In the human brain, just a bit behind the cell groups in the previous figure, Dick Swaab and his team in Amsterdam found that the bed nucleus of the stria terminalis (BNST) is more than twice as large in a man's brain as in a woman's. Swaab's paper and other interesting papers in this area of study can be seen in the journal *Hormones and Behavior*. The BNST is adjacent to a major nerve fiber group called the fornix (FX). Other abbreviations are as in the previous figure.

merely a neuroanatomical result of some other factor that happens to be correlated with the psychological change, is not yet known. Thus, we don't know yet whether such neuronal differences are correlated with how children with ambiguous genitalia, for example, should be treated in their gender-differentiated roles. All of these sex differences in human brain anatomy do, however, bear on arguments about how people view the sexual aspects of their own and others' lives.

Gender Role Assignment

If some of the forces driving gender identity depend on specific neuro-anatomical features of a person's brain, and how a person subsequently feels, other forces come from the advice of a variety of medical and mental health experts. Individuals with atypical sexual development have choices, now, and are attended to carefully. Consider another example of the kind of patients I discussed in Chapter 10—in this case, the Klinefelter Syndrome kids, with XXY (the most common sex chromosome disorder) having small testicles, breasts, and so forth. From a scientist's theoretical point of view, to quote clinical endocrinologist Louis Gooren, of the Free University of Amsterdam, "We are far away from any comprehensive understanding of hormonal imprinting on gender identity formation." But from a practical, medical point of view, Dr. Justine Schober, whose surgical work I mentioned in Chapter 10, says "Medicine is much less patriarchal than it used to be. From a surgeon's point of view, there is more emphasis on giving people what they need and, in a human sense, not more than they want." Thus, we are speaking not just about a person's genital anatomy, but about a person's gender identity—that person's deep psychological sense of being male or female, regardless of how he or she looks. Paraphrasing Dr. Robin Dea, a mental health director for the Kaiser Permanente program in California, "we feel our gender identity in our soul, but it is on a continuum and it can evolve." One biologist, the lesbian spokeswoman Professor Anne Fausto-Sterling, talks about the cultural components of homosexuality, and goes so far as to talk about "the body as a social construct." She favors halting sexual assignment surgeries on infants, because infants are too young to make their own choices. Other professionals disagree with this "free to be" approach. Dr. Kenneth Zucker, a mental health professional in Toronto, tries to help children be comfortable with their sex as determined by their chromosomes until they are old enough to make their own choices.

As a disinterested scientist, not a medical doctor, I cannot see the point of simply halting sex assignment surgeries for babies with ambiguous genitalia, as favored by Professor Fausto-Sterling. Instead, my bottom-line opinion follows that of the expert Heino Meyer-Bahlburg. Regarding each choice of treatment (and attitude) for each child with atypical sexual development, Meyer-Bahlburg looks for a consensus among the members of a multidisciplinary team: the parents, a pediatric urologist, a psychologist, a medical ethicist, perhaps a genetic counselor, a pediatric endocrinologist, and perhaps others. With respect to possible surgery, "gender assignment," sex hormone treatment, and the child's psychological development under the intense and continuing stress of a disorder of sexual development, all of the expertise of such a team will necessary to assure optimal care.

Following on from gender assignment are questions of what abilities the kid is usually expected, as a result of such assignment, to have or not to have.

Cognitive Differences and Intellectual Overlaps

Many authors who focus on sex differences in human behavior get carried away. Instead of admitting the huge overlaps between men and women in a wide range of cognitive abilities, they magnify small differences.

Famously, Lawrence Summers, during his tenure as president of Harvard University, hazarded the guess that women did not go into mathematical and scientific careers as frequently as men do, because they are not very good at math and science. Wrong! In the United States, male/female differences in math performance by grade school students are small and inconsistent, according to statistician Janet Hyde and her team at the University of Wisconsin, reporting data from the National Assessment of Educational Progress. Further, mathematicians Stephen Machin from University College, London, and Tuomas Pekkarinen from the Helsinki School of Economics, recently reported the results of an assessment among industrialized countries by the Organization for Economic Cooperation and Development (OECD). They wrote that in countries whose cultures tend to emphasize gender equality, "the differences in average mathematics test scores, usually in favor of boys, are erased or even reversed in favor of girls." Instead, they find that the variance among scores achieved by boys is reliably higher than the variance among scores achieved by girls. Thus, those genius boy mathematicians far to the right in their curve are, to some extent, balanced by other boys far to the left. For reading scores, it is the other way around. On the average, girls are indeed slightly better at reading, but the point about variance within a sex again comes into play. For reading, I once more quote Machin and Pekkarinen: boys' "higher variance in reading is due to a greater preponderance of boys in the bottom part of the test score distribution."

We have not yet considered the differences between innate abilities and environmental effects on possible sex differences. Some of those genius boy mathematicians will have been given systematic attention and

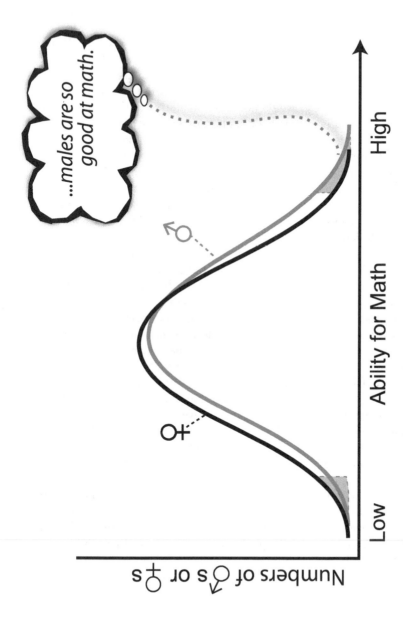

There is a tremendous overlap between males and females in mathematical ability. Particularly in countries with gender-equal cultures, the distribution of males' scores might encompass a wider range. Those at the top of the distribution (sticking out to the right in this drawing) are our "male math geniuses."

encouragement throughout. The opposite situation can obtain for girls, who may receive discouraging signals from the environment. My own daughter went to a premier high school in Scarsdale, New York. When she got the best score in an advanced mathematics class (that featured three girls and seventeen boys), the (superannuated male) teacher held up her test and asked "are we going to let a girl get the best score?" Now, if I were not her father, I would not know about this embarrassing event. My point is that it is extremely difficult to *rule out* environmental causes of sex differences in cognitive capacities, because to do so, you'd need comprehensive knowledge of kids' upbringings, day by day, year by year. To quote the Italian economist Luigi Guiso, leader of an international team who studied mathematical and reading performance, "social conditioning and gender-biased environments can have very large effects on test performance." In girls as well as boys, promoting math skills produces math skills.

Other cognitive abilities, having somewhat more obscure applicability to daily life, do show sex differences. In the words of Melissa Hines, "mental rotations, or the ability to rotate two or three-dimensional stimuli in the mind rapidly and accurately, shows a sex difference favoring males." On the other hand, tests of perceptual speed and perceptual accuracy yield significantly higher scores for females. Again, we don't know where these differences come from or exactly what they mean.

I agree with the evolutionary psychologist David Buss, from the University of Texas, when he says, "men and women differ more in mating than in any other area of life." That is, putting all we have discussed about brain function in perspective, the farther we get away from nitty-gritty reproductive function, the less impressive and reliable sex differences are. To put it another way, much more robust than tiny cognitive differences are sex differences in personality and attitude toward mate choice and gender role.

No one would say that men's and women's cognitive abilities and emotional expressions are absolutely identical—the old story still may have some truth: that women are more likely to place high value on a potential mate's financial capacity to support her and her children in the future. And it may still be true that men will look preferentially at a woman's youth and physical appearance, not only as signs of her fertility but also to raise his own social status. But overgeneralizing from these differences related to reproduction, to other areas of behavioral performance, would lead to big mistakes.

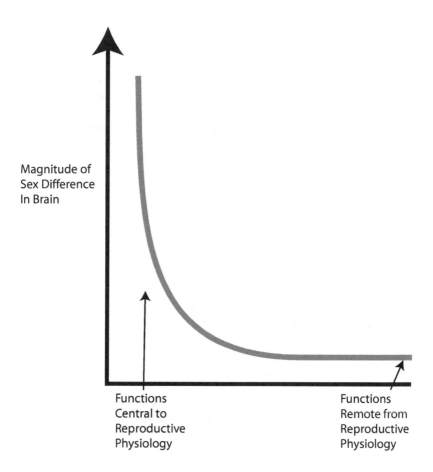

A sketch illustrating the idea that, even as the human brain features massive sex differences when it comes to the actual, biological regulation of sexual reproduction, the farther away from reproductive biology, in human experience and human behavior that we consider, the smaller and more variable any sex difference will be.

Man and Woman: Personalities and Mating Strategies Across the Continents

The evolutionary psychologist David Schmitt, of Bradley University, has been studying the personalities and mating habits of men and women from 56 nations spreading across 6 continents, 13 islands and 28 languages.

Several years ago he founded the International Sexuality Description Project for this purpose, with the idea of discovering potential predictors of sexual aggression, domestic violence, and exposure to risks for AIDS. Across the board, in his words, "women reported significantly higher levels of neuroticism, agreeableness, extraversion and conscientiousness than did men." Moreover, there were differences among nations in the degree of sex differences detected in these personality traits. Roughly speaking, nations characterized by longer healthy lives, equal access to education and economic wealth, reported larger sex differences in these aspects of personality. Schmitt speculates that in developed nations, expression of these particular aspects of personality is less constrained by circumstances, so that natural differences appear, whereas in less fortunate circumstances "innate personality differences between men and women may be attenuated."

Mate Poaching and Mate Guarding

During a woman's lifetime she will have four major types of hormonal fluctuations—during puberty, during her menstrual cycle, as a result of pregnancy, and during menopause. Social anthropologists have tended to focus on behavioral changes during the menstrual cycle, as they influence the woman's feelings. According to Steven Gangestad, anthropologist at the University of New Mexico, women tend to feel more physically attractive and to have more interest in meeting men as they approach their time of ovulation. Pair-bonded women reported "more extra-pair flirtation" and also, interestingly, at this time when the risk of cuckoldry was higher, more "mate guarding" by their primary partner—that is, more attentiveness, jealousy, possessiveness, and even preemptive intercourse by their partner. There were variations on a theme. If their primary partner was low in sexual attractiveness, these women particularly noticed mate guarding. Correspondingly, if the women were very attractive, they "experienced relatively high levels of mate guarding throughout their cycle." Of course, the flip side of mate guarding is mate poaching, the attempt by an outside male to attract another man's female—from the point of view of the attractive woman's mate, the poaching to be guarded against! David Schmitt reports that mate poaching by young adults is most common in Europe and the Americas, less frequent in Africa and Asia. Men do it more than women.

On the other hand, mating strategies show some sex differences that are universal across cultures. In all regions of the world sampled, Schmitt found that young adult males, on average, have a greater desire for sexual variety than do young adult females. As Schmitt wrote, "In humans, because men tend to be the lesser-investing parent of our species, they have more to gain than women do from indiscriminately engaging in short-term sex with numerous partners." As against this tendency, what keeps a couple together?

Is there an underlying human tendency for men and women to stay with the same mate for life? Schmitt reviews the evidence. One size definitely does not fit all. On the one hand, according to Schmitt, "most marriages in preindustrial cultures are socially monogamous," and several facts of human social life match those of lower primates that are indeed monogamous. On the other hand, other human characteristics are the same as those of lower primates whose species sport multiple pair bonds, or "polygyny": males with high status having multiple wives and frequent copulations, foragers living in large bands, human females being concentrated and guarded in foraging cultures, and vigorous competition among males for mates. We see polygynous mating strategies in which individual males mate with numerous females, while females mate only with one male, preferably a "high status man with ample resources." We seldom see polyandrous mating strategies in which a female will compete for access to numerous males, sometimes copulating with a male but then leaving him and the offspring.

Evolutionary thinkers try to account for this variety of mating patterns by resorting to all-encompassing theories. For example, "parental investment theory" concentrates on the differential time and energy fathers and mothers put into caring for offspring. Many men invest as heavily as women, but overall there is a much greater biological and psychological investment by mothers. The "lesser investing sex," men, are free to compete for other females, sometimes indiscriminately. David Buss and David Schmitt have proposed a "sexual strategies theory," which proposes that "men and women have evolved a complex repertoire of mating strategies." The balance between long-term and short-term mating, according to them, depends on a large number of environmental factors including "opportunity, personal mate value, sex ratio in the relevant mating pool, parental influences, and regnant cultural norms." For example, some scholars find that children who were raised in very stressful circumstances "tend to develop insecure parent-child relationships" and tend toward short-term mating. Conversely, children raised under supportive and comfortable conditions head eventually toward long-term mating. Variety among humans is the name of the game.

Consider, with the expert Columbia professor Anke Ehrhardt, all the behavior patterns we can measure: frank courtship and sex behaviors, yes, but also sexually dimorphic behaviors such as maternal behavior, gender role behaviors such as rough-and-tumble play, gender identity (self-awareness of oneself as male or female), and sexual orientation (including the choice of partner as a homosexual, heterosexual or bisexual).

All of these examples consider straight, heterosexual individuals. Where, in this great variety of choices about sexual behaviors and gender roles, does homosexuality come in?

A Person's Own Sexual Preference

At this point we enter the field of debate about the causes of homosexual feelings. How much do hormones have to do with homosexuality? On the one hand, there is clear evidence, mentioned earlier, that early exposures to physiological levels of androgenic hormones will dispose the individual toward a masculine gender role. Consider what happens, in the extreme case where these androgenic hormones are *not* synthesized or can not act normally. In XY males having mutations that disable an enzyme essential for the synthesis of testosterone, or that disable the androgen receptor, genital anatomy may not be masculinized and, accordingly, the young children are "assigned" as females and live their lives as females. These individuals give us an extreme example of a simple statement: "hormones and subsequent determination of genital anatomy can determine gender role." On the other hand, we must attend to the subtleties and complexities of gender role *choice*. Importantly, the nature of a child's social experiences, including closeness to one parent or the other, can have an effect on later psychosexual development. Little boys who feel they want to be girls, had peer groups of little girls. Little girls who stated their wish to be boys had primarily male peer groups and loved rough-and-tumble play.

It isn't just the presence of a given hormone that determines sexual preference and gender role choice. David Rubinow and Peter Schmidt, at the National Institute of Mental Health, point out that the influence of a given hormone on human behavior can depend on the *context* in which that hormone is operating. What other hormones are present? How old is the person? What time of day is it? Is the person stressed? What is the person's social rank and past social experience?" From laboratory experiments and clinical experience, Rubinow and Schmidt could show that contextual factors are important, and defeat the idea of "simple causal relations" between hormones and human behavior.

Homosexuality, according to Vivienne Cass, psychologist at the University of Western Australia, was initially viewed as a pathological state, a medical problem resulting from purely biological factors. Certainly, from studies of identical twins, some biologists have reached the conclusion that fifty percent or more of the causation of homosexuality is due to heredity. However, Cass also notes that Freud suggested a different type of cause—that homosexuality was a "natural process in the developing individual" and that if an individual "did not achieve the final phase of development, which was heterosexuality," then homosexuality might result. Freud's theory placed more emphasis on underlying psychological factors and relations within families. Along these lines, I am most amazed by the fraternal birth order effect. Paraphrasing Anthony Bogaert, psychologist from Brock University in Ontario, Canada, the most consistent correlate of sexual orientation in men is the number of older brothers: more brothers, more chance of homosexuality. While Bogaert mentions that these data "suggest a prenatal origin to the fraternal birth order effect," I suspect that things are much more complicated. I hold with Vivienne Cass in suspecting that "sexual behavior is not simply an internal quality but involves significant social, public and relational dimensions." Envisioning the fraternal birth order effect from this perspective leads to the view that the social relations of a young boy with his (many) older brothers almost certainly will influence his sexual orientation. He is not the biggest guy in the family. He may be the smallest, and completely unable to assume, or even to imagine himself as assuming a masculine gender role and a heterosexual identity.

Eric Vilain, at UCLA Medical School, reminds us of how complex the genetic contributions to sex determination can be. As discussed in Chapter 2, SRY is an excellent candidate on the Y gene to determine the testes: its conservation across mammalian species, its expression in the genital ridges just prior to male genital tract experience, and the ability of a single base change in the SRY region coding for its protein to, in Vilain's words, "change the fate of the bipotential gonad of an XY fetus from testicular to ovarian development." However, only about 25% of females with XY chromosomes have mutations in the SRY gene. There is a lot more going on, even in the biological chain of events leading to gender role and associated sexual preference.

In this field of science and medicine, it is terribly easy to oversimplify. So very many variations can be cited. Consider the famous composer, pianist and teacher Leonard Bernstein, the late conductor of the New York Philharmonic Orchestra. He married the beautiful Costa Rican actress Felicia Montealegre Cohn, an actress whose child he fathered. Yet, throughout his adulthood, he had many intense homosexual affairs. The biographer Humphrey Burton speculates that Felicia married Bernstein "knowing that he was bisexual"! What you would expect from publicly available information differed so markedly from his libidinous impulses. With respect to the complexities of Bernstein's sexual preferences, as contrasted to his marital life, I can cover the same type of comparison in a less biographical and more scientific manner. Peter Todd and his associates at the Max Planck Institute for Human Development in Berlin analyzed statistical data from self-reports before, during, and after speed dating sessions. They came to the clear conclusion that the cognitive processes that drive mate preferences are not the same as the cognitive processes that underlie eventual mate choice. When asked about preferences with respect to qualities like wealth, status, family commitment, physical appearance, overall attractiveness, and healthiness, both men and women followed the "likes attract each other" law. People preferred potential mates who matched their perceptions of themselves. But these stated preferences for mates did not match what the subjects actually did. Men chose women based, purely and simply, on their physical attractiveness. Women, who generally were more discriminating, chose men whose overall qualities as a permanent mate, with respect to all of the qualities mentioned, matched the women's own self-perceived physical attractiveness. These results followed absolutely from the evolutionary theory of mate choice according to the nature of parental investments. That is, the authors felt that their results shed light on "how evolution has biased the male and female mind in different directions." And the idea of "choosy females and competitive males" fits well with Darwinian evolutionary thinking.

How the Story Ends

At this point we have identified a large number of social, psychological and biological influences on individuals' choices of gender roles. Is there a picturesque way of putting them all together and making a conclusion? Perhaps they are best visualized by taking off from a classic illustration used by the great British developmental biologist C.H. Waddington, in his 1957 classic, *The Strategy of the Genes*. Picture a mountain in the winter with ski trails. In the illustration that follows, just a few of the influences on gender role are pictured as intersections among trails. Thus, the baby could have two X chromosomes or could have an X and a Y chromosome. X chromosome inactivation could happen this way or that way. Testosterone (TST) and anti-Müllerian hormone (AMH) might or might not reach the brain in high concentrations. Later on, for this same child, puberty might be very stressful, or not. And when this baby has reached young adulthood, its town might have attractive potential mates of the opposite sex, or maybe not. The old view encompassed only strongly masculine or strongly feminine gender identifications. Now, in modern society, we can understand that there are many, many ways of arriving at a large number of points at the bottom of the mountain. That is, the very multiplicity of causes of sexual orientation, and the graduations of levels of each cause, lead me to conclude that there is tremendous flexibility in a person's assumptions of sex roles, and that they are reversible. Further, they are not monolithic and dichotomous. A person could happily participate in a typically male role in some aspects of life, and typically female in other. These are the true facts, especially in modern societies where large muscle strength is not as much a premium, and where women are effective as they go to war.

I would not go so far as the well-known psychoanalyst Nancy Chodorow, who has described "gender as a personal and cultural construction." After all, Chodorow opens her own book, *The Reproduction of Mothering*, with the simple declarative sentence "Women mother." And she follows that up by stating, "women by and large want to mother, and get gratification from their mothering; and finally...have succeeded at mothering." I do not believe, as some do, that it is merely a habit to think of other individuals as men or women. In humans' use of language, the verbal distinction between male and female seems to be universal (refer back to the first figure in Chapter 1). Nevertheless, in the words

OLD. For the developing child (pictured on top of a metaphorical mountain), strongly differentiated masculine and feminine gender roles were primarily available.

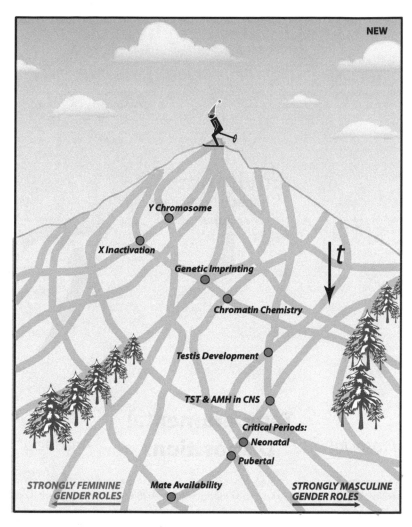

NEW. In many modern societies, a wide variety of flexible and nondichotomous (intermediate, mixed) gender roles are experienced and practiced. The wide variety is produced by the large number of genetic, hormonal and environmental (especially neonatal and pubertal) determinants, a few of which are sketched here. Sexual differentiation of human behavior offers one illustration of how gene/environment interactions take place. You can see this as you follow the routes the modern child can take from the top to the bottom of the "mountain."

Genes influencing behaviors

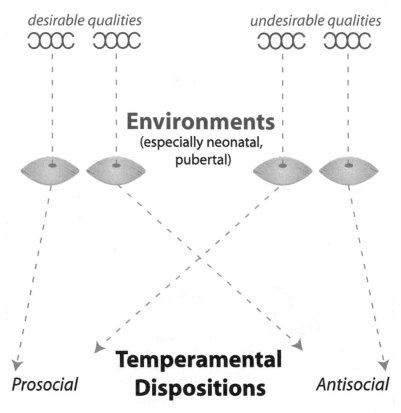

desirable qualities

undesirable qualities

Environments
(especially neonatal,
pubertal)

Prosocial

**Temperamental
Dispositions**

Antisocial

A cartoon that uses an optical lens analogy to illustrate how favorable environments can transform behavioral tendencies emanating from less-than-fortunate genetic constitutions (right side of sketch) toward more friendly, prosocial dispositions. In the opposite direction, terrible environments can degrade the initial behavioral tendencies of a child with a great inheritance (left side of sketch), leading to antisocial temperamental qualities. The cartoon pictures four extreme types of gene/environment interactions.

of Professor Anne Fausto-Sterling, "people come in bewildering sexual varieties."

This message carries huge implications for women's roles in the societies on this planet. We already know that Margaret Thatcher was a tough prime minister in the United Kingdom, that Golda Meir was a flame-throwing leader in Israel, and that Hilary Clinton can trump almost all American politicians. And, in recent years, the CEOs of the Hewlett-Packard, Lucent and Xerox corporations are, or have been, women. But writing in *Harvard Magazine*, the psychologist Harbour Fraser Hodder looks to the future and makes a deeper point. He talks about "alpha girls." He notes that "58% of kids who enroll in college are women" —in 1970, 58% were men. Interviewing high-achieving high school girls in the U.S. and Canada, he found some who "had attained a 3.8 or better grade point average and at least one leadership position, pursued 10 or more hours of extracurricular activities weekly, and scored high on measures of achievement motivation." These are alpha-girls. They "invest in their own human capital." Further on in life, after they have finished in college, fertility control will help women to balance the challenges of work and family in a manner more permissive of big careers than ever before. Looking back, I think that Thatcher, Meir and Clinton had to be tough in order to succeed in a male-dominated, competitive environment. In the future, with society ready for them, alpha-girls are going to have wide choices of leadership style. They will do good and create value, using a variety of leadership strategies not available to their mothers and grandmothers.

Finally, throughout this book, I have tried to attack oversimplifications. Don't ever think that a cause is "genetic *versus* environmental" as it influences a sex difference in behavior. Instead, it is "genetic *modified by* environmental" influences. As you can guess, for sex differences in animals, and especially in humans, we in the laboratory are still enmeshed in the task of figuring out exactly how men and women come to differ, when they really do differ in their behaviors, their hearts, and their minds. To take off from the words of David Brooks, in his column in the *New York Times* regarding genetic and environmental influences, we are still addressing "the ineffable mystery of why people do what they do."

Further Reading

As a scientist passionate about his fields of work, I have enjoyed telling this story. Really, two stories: one, the saga of developments in the body leading to sexually relevant behaviors, and two, the story of the most prominent scientists who have spent or are spending their lives discovering these facts. I have not tried to document every statement with references, although I wrote this book in my home study surrounded by piles of papers. This book had to be "reader friendly" in its writing and in its appearance on the page. To add every relevant reference would have overburdened the page by literally hundreds of footnotes that would have made my book seem more like a diatribe or a text than a celebration of scholarly efforts of current interest. So, instead of documenting every statement, I had each chapter vetted by experts in the field. Their efforts were much appreciated and are acknowledged, chapter by chapter, below.

I have chosen further reading with two criteria in mind. First, they should illustrate a main point described in the chapters for which they are listed. Second, they should be written in such a way that they are accessible to nonscientists educated through college.

For readers who want to go further into the nitty-gritty, the detailed accounts of primary data that support the story I have told, there are many sources easily found through Amazon and other routes. For aggressive behaviors, look for books by Randy Nelson. For maternal behaviors, Michael Numan. For sex chromosomes, writing by David Bainbridge. For evolutionary approaches to the behaviors recounted here, try David Buss. I, myself, have written about mechanisms of sex behavior and of brain arousal. For gender role assignments and choices, books by John Money and by Heino Meyer-Bahlberg. For teaching kids how to overcome cognitive or emotional limitations peculiar to their sex, books by Lisa Eliot. For coverage of the neuroscience of this entire field, I and a wonderful team of expert co-editors have edited *Hormones, Brain and Behavior*, a 5-volume series whose second edition has just appeared. And for the most detailed accounts of studies in the fields of work covered by this book, go to the Annual Reviews of Physiology, Psychology and Neuroscience. For access to the scientific literature beyond these sources, there is an automated, free search engine called PubMed.

Further Reading

Chapter 1: What Scientists Fight About When They Fight About Sex (Pages 3–16)

Pfaff, D. (2006). *Brain arousal*. Cambridge: Harvard University Press.
Rogers, L. (2001). *Sexing the brain*. New York: Columbia University Press.
Shipman, C. & Kay, K. (2009). *Womenomics*. New York: HarperCollins.

Further Reading

Chapter 2: Chromosomes for Him and Her (Pages 17–30)

Arnold, A.P. (2004). *Nature Reviews Neuroscience, 5*:701–708.
Carrel, L. & Willard, H. (2005). *Nature, 434*:400–404.
Koopman, P., et al. (1991). *Nature, 351*:117–121.
Kow, L-M., et al. (2007). *Current Topics in Developmental Biology,* 79:37–59.
Li, L-L., et al. (1999). *Science, 284*:330–333.
Maxson, S. (1996). *Behavior Genetics 26*: 471– 476.
Pfaff, D. (1973). *Science, 182*:1148–1149.
Schwanzel-Fukuda, M. & Pfaff, D. (1989) *Nature, 338*:161–164.
Sinclair, et al. (1990). *Nature, 346*:240–244.

Swaney et al. (2007). *Proceedings of the National Academy of Sciences,* *104*:6084–6089.

Further Reading

Chapter 3: Hormones on the Brain (Pages 31–50)

Auger, A., et al. (2000). *Proceedings of the National Academy of Sciences,* 97:7551–7555.

Bakker, J. & Baum, M. (2007). *Frontiers in Neuroendocrinology*

Gorski, R. & Barraclough, C. (1963). *Endocrinology, 73*:210–216.

Meisel R. & Ward I. (1991). *Science, 155*: 239–242.

Olsen K. (1979). *Nature, 279*:238–239.

Pfaff, D. (1980). *Estrogens and brain function.* New York: Springer Verlag.

Phoenix, C., et al. (1959). *Endocrinology, 65*:369–382.

Sato, T., et al. (2004). *Proceedings of the National Academy of Sciences,* *101*:1673–1678.

Sisk, C. & Zehr, J. (2005). *Frontiers in Neuroendocrinology, 26*:163–174.

Wagner, C. (2006). *Frontiers in Neuroendocrinology, 27*:340–359.

Wallen, K. & Baum, M. (2002). In: *Hormones, brain and behavior.* Academic Press/Elsevier, Chapter 69, pp. 385–445.

Zhou J., et al. (2005). *Proceedings of the National Academy of Sciences,* *102*:14907–14912.

Further Reading

Chapter 4: Neonatal Hormones, Brain Structure, and Brain Chemistry (Pages 51–68)

Auger, A., Tetel, M. & McCarthy, M. (2000). *Proceedings of the National Academy of Sciences, 97*:7551–7555.

Devidze, N., Mong, J.A., Jasnow, A.M., Kow, L.M., & Pfaff, D.W. (2005). *Proceedings of the National Academy of Sciences, 102*(40):14446–14451

McCarthy, M. (2008). *Physiological Reviews, 88*:91–134.

McCarthy, M., Auger, A. & Perrot-Sinal, T. (2002). *Trends in Neuroscience, 25*:307–312.

Mong, J., Kurzweil, R., Davis, A., Rocca, M. & McCarthy, M. (1996). *Hormones and Behavior, 30*: 553–562.

Oro, A., Simerly, R. and Swanson, L. (1988). *Neuroendocrinology,* 47:225–235.

Rhodes, M. & Rubin, R. (1999). *Brain Research Reviews, 30*:135–152.

Romano, G.J., Krust, A. & Pfaff, D.W. *Molecular Endocrinology,* 3:1295–1300, 1989.
Romano, G.J., Mobbs, C.V., Lauber, A., Howells, R.D. & Pfaff, D.W. (1990). *Brain Research,* 536:63–68.
Sickel, M. & McCarthy, M. (2000). *Journal of Neuroendocrinology,* 12:397–402.
Simerly, R. (2000). In A. Matsumoto (Ed.). *Sexual differentiation of the brain.* London: CRC Press, pp. 175–202.
Simerly, R. & Swanson, L. (1987). *Proceedings of the National Academy of Sciences,* 84:2087–2091.
Vasudevan, N. & Pfaff, D.W. (2007). *Endocrine Reviews,* 28(1):1–19.

Further Reading

Chapter 5: Mating and Parenting (Pages 69–92)

Adkins-Regan, E. (2005). *Hormones and animal social behavior.* Princeton: Princeton University Press.
Balthazart, J. & Ball, G. (2007). *Frontiers in Neuroendocrinology,* 28:161–178.
Hull, E., et al. (1995). *Journal of Neuroscience,* 15:7465–7471.
Hull, E., et al. (2002). In D. Pfaff, et al. (Eds.). *Hormones, brain and behavior.* San Diego: Academic Press/Elsevier.
Numan, M. & Insel, T. (2003). *The neurobiology of parental behavior.* Berlin: Springer-Verlag.
Ogawa S., et al. (1996). *Neuroendocrinology,* 64:467–470.
Ogawa S., et al. (2004). *Annals of the New York Academy of Sciences,* Vol. 2036.
Pfaff, D. (1980). *Estrogens and brain function.* New York: Springer-Verlag.
Pfaff, D. (1999). *Drive.* Cambridge: MIT Press.
Romano, G., et al. (1990). *Brain Research,* 536:63–68.
Wallen, K. (2001). *Hormones and Behavior,* 40:339–357.

Further Reading

Chapter 6: Males Fighting (Pages 93–114)

Devine, J., et al. (2005). *Scientific approaches to youth violence prevention. Annals of the New York Academy of Sciences,* Vol. 1036.
Nelson, R. (Ed.). (2006). *The neurobiology of aggression.* New York: Oxford University Press.
Ogawa S., et al. (1996). *Neuroendocrinology,* 64:467–470.
Ogawa S., et al. (1998). *Endocrinology.*

Pfaff, D. (2007). *The neuroscience of fair play.* Washington: The Dana Press.

Further Reading

Chapter 7: Females Befriending (Males, Too) (Pages 115–130)

Baron-Cohen, S., et al. (2004). *Prenatal Testosterone in Mind.* Cambridge: MIT Press
Choleris, E., et al. (2006). *Genes, Brain and Behavior,* 5:528–539.
Keverne, E.B., et al. (1999). In C.S. Carter et al. (Eds.). *The integrative neurobiology of affiliation.* Cambridge: MIT Press.
Pfaff, D.W. (2007). *The neuroscience of fair play.* University of Chicago: Dana Press.
Taylor, S., et al. (2000). *Psychological Review,* 107:411–429

Further Reading

Chapter 8: Pain and Suffering (Pages 131–150)

Altemus, M. & Epstein, L. (2008). In Becker, J. et al. (Eds.). *Sex differences in the brain.* New York: Oxford University Press.
Bale, T. et al. (2002). *Journal of Neuroscience,* 22:193–199.
Bodnar, R. et al. (2002). *Central neural states relating sex and pain.* Baltimore: Johns Hopkins University Press.
Goel, N. & Bale, T. (2007). *Endocrinology,* 148:4585–4591.
Murphy, A.Z. et al. (1999). *Hormones and Behavior,* 36:98–108.
Pfaff, D. (2006). *Brain arousal and information theory.* Cambridge: Harvard University Press.
Swanson, J. et al. (2008). *Journal of Attention Disorders,* 12:4–14; 15–43.
Young, E. & Libezon, I. (2002). In Pfaff, D. et al. (Eds.). *Hormones, brain and behavior.* San Diego: Academic Press/Elsevier.

Further Reading

Chapter 9: Perilous Times—Newborns and Adolescents (Pages 151–170)

Devine, J. et al. (Eds.). (2004). Youth violence: Scientific approaches to prevention. *Annals of the New York Academy of Sciences,* Vol. 1036.
Mueller, B. & Bale, T. (2008). *Journal of Neuroscience,* 28:9055–9065.
Harlow, H. (1974). *Learning to love.* New York: Aronson.
Sisk, C. & Zehr, J. (2005). *Frontiers in Neuroendocrinology,* 26:163–174.

Styne, D. & Grumbach, M. (2002). In Pfaff, D. et al. *Hormones, brain and behavior,* *Vol. 4.* San Diego: Academic Press/Elsevier, pp. 661–716.

Further Reading

Chapter 10: "Sex Gone Wrong" (Pages 171–186)

Mathews, G., et al. (2009). *Hormones and Behavior,* 55:285–291.

Meyer-Bahlburg, H., et al. (1996). *Hormones and Behavior,* 30:319–332.

Meyer-Bahlburg, H., et al. (2002). In Zderic, S., et al. (Eds.) *Pediatric gender assignment: A critical reappraisal.* Kluwer Academic/Plenum Publishers, pp. 199–223.

McGinnis, M., et al. (2002). In Pfaff, D. et al. (Eds.). *Hormones, brain and behavior,* *Vol. 5,* pp. 347–380.

Schober, J., et al. (2004). *British Journal of Urology International,* 94:589–594.

Woodhouse, CRJ. (2004). In Legato, M. (Ed.). *Principles of gender-specific medicine,* Vol. 1. San Diego: Academic Press/Elsevier.

Wunsch, L. & Schober, J. (2007). *Best Practice and Research in Clinical Endocrinology and Metabolism,* 21:367–380.

Zhu, Y-S & Imperato-McGinley, J. (2002). In Pfaff, D. et al., (Eds.). *Hormones, brain and behavior.* San Diego: Academic Press/Elsevier.

Further Reading

Chapter 11: Bottom Line (Pages 187–208)

Baum, M.J. (2006). *Hormones and Behavior,* 50:579–588.

Chodorow, N. (1978). *The reproduction of mothering.* Berkeley: University of California Press.

Chodorow, N. (1999). *The power of feelings.* New Haven: Yale University Press.

Gooren, L. (2006). *Hormones and Behavior,* 50:589–601.

Haselton, M. & Gangestad, S. (2006). *Hormones and Behavior,* 49:509–518.

Hines, M. (2002). In Pfaff, D., et al. (Eds.). *Hormones, brain and behavior.* 1st edition. San Diego: Academic Press/Elsevier.

McCarthy, M.M. & Konkle, A.T.M. (2005). *Frontiers in Neuroendocrinology,* 26:85–102.

Schmitt, D. (2005). *Behavioral and Brain Sciences,* 28:247–311.

Swaab, D., et al. (2001). *Hormones and Behavior,* 40:93–98.

Todd, P. (2007). *Proceedings of the National Academy of Sciences,* 104:15011–15016.

Acknowledgments

I would like to acknowledge the members of my laboratory at the Rockefeller University who have worked so hard to contribute data to our part of this story. Marion Osmun and Craig Panner, my editors at Oxford University Press, were encouraging throughout, even as they gave me precise and useful critical advice. Two of my previous books, *Brain Arousal* and *Neuroscience of Fair Play* gave me a running start on the concepts presented here, and I thank my editors who helped me with those, as well. Susan Strider, at Rockfeller University, provided all of the illustrations. Luba Vikhanski, of the Weizmann Institute, gave me the title, and some wording in Chapter 1.

Chapter 2: William Crowley, MD, Professor at Harvard Medical School; E. Barry Keverne, PhD, Professor and Chair of Animal Behaviour at the University of Cambridge; and Arthur Arnold, Professor at UCLA, were kind enough to read and comment on Chapter 2.

Chapter 3: Professors Michael Baum, Boston University; Roger Gorski, UCLA; Diane Robins, University of Michigan; Cheryl Sisk, Michigan State University; Kim Wallen, Emory University.

Chapter 4: Professors Margaret McCarthy, University of Maryland medical school; Dick Swaab, MD, Director Emeritus, National Institute for Brain Research, Amsterdam; Professor Richard Simerly, University of Southern California.

Chapter 5: Professors Elizabeth Adkins-Regan, Cornell University; Elaine Hull, Florida State University; Michael Numan, Boston College.

Chapter 6: Professors Bob and Caroline Blanchard, University of Hawaii; Edward Brodkin, University of Pennsylvania; David Edwards, Emory University; Randy Nelson, Ohio State University; Michael Potegal, University of Wisconsin.

Chapter 7: Professors Carol Sue Carter, University of Illinois at Chicago; Helen Fisher, PhD, Rutgers University; and Shelley Taylor, UCLA.

Chapter 8: Professor Margaret Altemus, MD, Weill Cornell Medical College; Professor Tracy Bale, University of Pennsylvania; Karen Berkeley, Florida State University; Rebecca Craft, Washington State University; Elizabeth Young, University of Michigan.

Chapter 9: Professor Frances Champagne, Columbia University; Professor Melvin Grumbach, UCSF; Professor Bruce McEwen, Rockefeller University.

Chapter 10: Justine Schober, M.D., Hamot Medical Center and Rockefeller University.

Chapter 11: Professors Anne Etgen, Albert Einstein College of Medicine; Margaret McCarthy, University of Maryland; Bruce McEwen, Rockefeller University; and Dick Swaab, Netherlands Institute for Brain Research.

Index

Meyer-Bahlburg, Heino F. L., 176–77, 180, 182, 193
Micropenis, 174–76
Migraine headache, 132
Mill Hill, National Institute for Medical Research, London, 20
MIS. *See* Müllerian inhibiting substance
MIT. *See* Massachusetts Institute of Technology
Moir, Anne, 7–8
Molecular endocrinology, 51–53, 52*f*
Mong, Jessica, 41
Monogamy, 116, 124, 199
Mothers, 48, 70*f*, 88–92, 91*f*, 92*f*, 182, 191, 203, 205
  aggressive behavior in, 113
  maternal love, 151–55
  as model for all affiliative behaviors, 87
  parental "imprinting," 23–24
  "parental investment theory," 199
MRNA. *See* Messenger RNA
Mueller, Bridget, 153
Müllerian inhibiting substance (MIS), 25–27, 25*f*
Multiple chemical sensitivity (MCS), 145–46, 145*f*
Murphy, Anne, 134–35, 135*f*
Musatov, Sergei, 137–38, 138*f*
Muscles, 61, 95, 97, 132, 140, 143, 154, 180, 203
  growth in males, 70*f*
  joint/muscle pain, 146*f*
  lordosis, 82*f*
Mutation, 7, 20, 38, 103, 179, 180, 181*f*, 182, 186, 200, 202

National Assessment of Educational Progress, 194
National Institute for Brain Research, Netherlands, 58, 190–91, 191*f*
National Institute of Mental Health, Maryland, 87, 132, 200–201
National Science Foundation, 37
Natural selection, 6, 70–71, 121, 168
Nausea, 145–46, 145*f*
Neonatal period, 5, 37, 41–57, 54*f*, 151–53. *See also* Parenting behaviors
  glia gene expression in males *v.* females, 59–66, 60*f*, 62*f*, 63*f*
Nerve cells, 42*f*
  gap junctions, 41
  sex hormones influencing, 40*f*
  tunnel formation between, 42*f*
Netherlands, 58, 97, 190
Neurochemicals

directly involved in aggressive behaviors, 102–6, 105*f*
for labor and delivery pain, 132–34
The Neuroscience of Fair Play (Pfaff), 116, 123, 125, 168–69
Neurospora fungus, 120
Nippon Medical School, Tokyo, 65
Nitric oxide synthase, 65
Norepinephrine, 128, 139*f*, 141, 142*f*
Nose. *See* Olfactory cues
Nuclear envelope, 40*f*
Numan, Michael, 87–92
Nutrition
  ER-beta estrogen receptor, 186
  onset of puberty and, 161–63, 162*f*
  ventromedial nucleus role, 131, 136–37, 138*f*

OECD. *See* Organization for Economic Cooperation and Development
Ogawa, Sonoko, 102*f*
Ojeda, Sergio, 48–49, 63*f*, 163
Olfactory cues, 28*f*, 70*f*, 75
  MCS and CFIDS onset, 145–46, 145*f*
  social recognition from, 116–21, 117*f*, 119*f*
Olsen, Kathy, 37
"One gene/one enzyme" principle, 8, 120, 202, 207
Ong, Albert, 174
Opioid gene systems, 61, 134–35, 135*f*
Optical lens analogy, 206*f*
Optic chiasm, 36*f*, 191*f*
"Optimal gender" *v.* "true sex," 182
Organization for Economic Cooperation and Development (OECD), 194
OT. *See* Oxytocin
OTR. *See* Oxytocin receptors
Ottinger, Mary Ann, 185
Ovarian estrogens, 46–47
Ovaries, 24–25, 25*f*, 36*f*, 39, 49
  maternal behavior, 88–92, 91*f*, 92*f*
  onset of puberty, 162*f*
Oversimplification ("one gene/one behavior"), 8, 120, 202, 207
Ovulation, 33–35, 34*f*, 36*f*, 46–47, 49, 64, 86*f*, 198
  preoptic area's role, 56, 65
Oxytocin (OT), 156
  in social recognition mechanisms, 118–19, 119*f*
Oxytocin receptors (OTR), 65, 88–92, 91*f*, 92*f*

Pain, reception and suppression of, 132–36, 133*f*, 135*f*. *See also* Female suffering
  male suffering, 146–48, 146*f*, 148*f*